The End of the World or a New Era?
Eight World Scenarios of Death and Life
(Edición 1.0)

Ambrose Goikoetxea, Ph.D.

agoikoetxea1@telefonica.net

15 May 2017

Ambrose Goikoetxea, Ph.D.

Published by:
Euskal Herria 21st Century Press
Ave. La Rioja 15, Laguardia 01300, Alava, Basque Country
Tel: 628 70 36 16
www.euskalherriasiglo21.org

The End of the World or a New Era?:
Eight War Scenarios of Death or Life
(Edición 1.0)

ISBN-13:   978-1546890591
Deposit legal: (In progress)
Printed by CreateSpace, USA (www.createspace.com)

Table of Contents

Chapter 8: _Scenario 4_: _Nuclear Warfare and death of the Human Genome_

Chapter 9: _Scenario 5_: _Diversity of Religions and Beliefs_

Chapter 10: _Scenario 6_: _Promote Economic and Political Balance across the Global Community_

PEOPLE AND PLACES

- *Kathy Thompson*, is a young American woman from Tucson, Arizona, USA, 26 years old, half Anglo, half Jewish, and one-hundred percent intelligent and beautiful. A researcher of *Gnostic codes* at the University of Arizona where she studies to earn a degree and make a living in a society dominated by men, mostly. She's independent, sure of herself, ready to tackle most challenges in life, and not willing to limit herself to men's view of the world, until one day she meets Xabier Elurmendi, her life takes on new meaning, and together they venture into romance, corruption in high places, and a mission to save the Pope's life..

- *Xabier Elurmendi,* is a young man in the town of Bergara, Gipuzkoa, Basque Country, 32 years old, beginning a new career as an *Ertzaina* (Basque policeman). He gets involved in a bus accident on a mountain road carrying a group of young men and women that earlier had participated in a peaceful demonstration to "stop the TAV", a high-speed train that would run across the Basque Country. The bus slides down a mountain, several people get killed including friends of his. Impacted by this experience, Xabier joins the Arantzazu Seminary, a Franciscan seminary, to become a priest one day, to redirect his life. His interest for the history and life of the first Christians takes him on a one-year trip to the USA where he meets the very attractive and intelligent Kathy Thompson. There, his new faith, love, and carnal desires compete for his soul and body.

- *Dr. Eugene Finley,* is a professor, mentor of Kathy Thompson, and chief of a team of investigators at the University of Arizona, Tucson, Arizona. This team is working with *the International World Organization for Peace (IWP)* in the design of the 8 scenarios presented in this book.

INTERNATIONAL ORGANIZATIONS
AND TREATIES

1. *Treaty of Paris (2015)*, an international agreement regarding global warming: https://en.wikipedia.org/wiki/Treaty_of_Paris

2. *Earth Day,* an annual event celebrated on April 22 of each year, to demonstrate support for environmental protection: https://en.wikipedia.org/wiki/Environmental_protection

3. *Kyoto Protocol,* an international treaty which extends de 1992 United Nations Framework Convention on Climate Change (UNFCCC), which commits nations to reduce greenhouse gas emissions: https://en.wikipedia.org/wiki/Kyoto_Protocol

4. *List of International Environmental Agreements*, mostly legally binding for countries which have formally ratified them: https://en.wikipedia.org/wiki/List_of_international_environmental_agreements

5. *National Environmental Policy Act of 1969*, a United States environmental law to promote the enhancement of the environment: https://en.wikipedia.org/wiki/National_Environmental_Policy_Act

6. *Treaty of Non-Proliferation of Nuclear Weapons (NPT)*, an international treaty to prevent the spread of nuclear weapons and technology, and to achieve nuclear disarmament, opened for signature in 1968: https://en.wikipedia.org/wiki/Treaty_on_the_Non-Proliferation_of_Nuclear_Weapons

7. *Treaty for the Prohibition of Nuclear Weapons in Latin America and the Caribbean (Treaty of Tlatelolco),* signed in 1967 in the City of Mexico: https://en.wikipedia.org/wiki/Treaty_of_Tlatelolco

8. ***Principles Governing the Activities of States in the Exploration and use of the Outer Space,*** a treaty signed in 1967 in London, Moscow, and Washington D.C. by the United Kingdom, Northern Ireland, the Soviet Union, and the USA: https://en.wikipedia.org/wiki/Outer_Space_Treaty

Dedication

To Sofía, Ava, Eva and David.
To Esperanza ("Espe"), Javier y Andrea.
To Miguel, Jennifer, Marta, Rick, Charlie y Juliane.
To my parents, Teresa y Eusebio.
Mother Superior Esperanza Martinez.
To Jesus Blanco, Maria Kraus, Cristina, and brothers.
To Enrique Von Borstel.

To **men and women** of all ages interested in learning where we come from, learning about the evolutionary processes of life, our finite life in the planet and in the Universe, and the reasons for wanting to go on, to live as a species, **to enjoy life and scientific discovery** in our planet, now, and never again.

Very specially to my friends and colleagues in **Iniciativa Atea** (www.iatea.org), **American Atheist Center** (www.atheists.org/about-us) and other Atheist and Humanist organizations in the USA and worldwide for their dedication and effort on behalf of truth, understanding, social justice, and service to the community.

Preface

In this book I wanted to write about ideas and facts which have been flying inside my head for many years. Ideas and facts about the origins of our planet Earth, the origins of our human species, *Homo Sapiens Sapiens*, its evolution through millions of years, the creation of our cultures and civilizations, our exodus of some 300 families from the *African continent* some 50,000 years ago on our way to populate the other continents. The vast majority of this knowledge is due to the men and women in our history dedicated to research, to search for the origins of manking, the creation of technologies, political frameworks, and the creation of social models with which to build our towns and cities around the world, on planet Earth.

To that end, I have designed and described *8 scenarios* to try to explain where we are today in our planet with its 7,000 Million people. The first 4 scenarios I call *"fight-and-die" scenarios* because they illustrate where we are going today as a human species in a world where overpopulation, climate change, limited energy resources, religious wars, and an abundance of nuclear warheads threaten our own existence. The next 4 scenarios I call *"Negotiate-and-live" scenarios* because they illustrate how we might be able to change the course of the first 4 scenarios through agreements and respect for diversity of religions and beliefs, a better understanding of our human nature, the exploration of other planets in our own galaxy in which to create and sustain space colonies, and the promotion of an economic and political balance among nations in the world. *What is the likelihood of these scenarios happening?* Well, some of these scenarios are already happening, they are here with us today. The irony of it all is that the first 4 scenarios are more likely to happen and they would be amply deadly, in part because they make use of highly destructive technologies and religious beliefs intended to deny the common origins of all our ethnic groups and cultures. The last 4 scenarios represent a ray of hope to our humanity, but they are less likely to happen because they would require our many communities with their social and political ideologies around the world to recognize common attributes and origins in an effort to create a new world order based on economic and political balance, our origins in the Universe. These scenarios are *not "mutually exclusive"* and, as such, more than one of these scenarios could all be happening at the same time.

I chose three main characters, *Kathy Thompson, and Dr. Eugene Finley*, to play the role of detailing each one of these 8 scenarios, by means of questions and answers at a Seminar Series held at the *University of Arizona (UofA)*, in Tucson, Arizona, USA.

Accordingly, the *first 4 chapters* present a review of our current technologies, our current thinking and knowledge as to the origins of our human species, the list of wars we have engaged into over the centuries, and views on some of the apocalyptic social models in our immediate future unless we change, unless we consider altering our ways of doing things on planet Earth.

In *Scenario 1*, in Chapter 5, "*Slow Death by Overpopulation, Food Shortages, and Internal Conflict*", the main factors contributing to this scenario are listed and amplified, including overpopulation, food shortages, list of countries leasing agricultural lands in Africa and Australia, the emergence of criminal road mafias, and collapse of the world order. A long list of internal and guerrilla war activities are likely to emerge across the globe, listing possible international wars among countries like Syria, Russia, USA, Colombia, Mexico, Germany and Japan to mention only a few. Criminal mafia networks would exploit refugee mafias fleeing one conflict in their own countries and searching for a new life in other countries and continents. In the middle of all this chaos, the

International World Organization for Peace (IWP) works to coordinate flow of food resources, control crime, and promote political and military agreements among the many countries involved.

We are running out of oil resources in our planet. It is in Chapter 7, **Scenario 2.** *"Control of Energy Sources, Guerrilla Warfare across the Planet",* where we explore the events which will most likely happen as our oil reserves vanish across the planet, particularly in countries like Saudi Arabia, Iran, Irak, USA, Kuwait, Venezuela, and United Arab Emirates. Included in the list of main factors are to be found declining global economies, price increase in remaining oil resources, warm temperatures throughout the planet, international weapon supply mafias, and collapse of current world order. *Will alternative energy sources be in place to replace the vanishing oil resources?* Very likely we will see wars with allies such as Iran and Irak against combined warfare activity by USA, England, France, Israel, and Spain. By then our world population may have increased to 9,000 Million people, complicating nation's co-existence considerably.

Are there still non-believers of the climate change (CC) coming up in the very near future? In fact, it has already arrived as we show in great detail in Chapter 7, **Scenario 3,** *"Climate Change and Environmental Holocaust",* with a long list of main factors, including a death toll of 300,000 people per year, hunger affecting some 45 Million people in several continents, costs of CC in the order or 125,000 $Million/year, and the threat to 1/4 of the flora in our planet. In particular, we take a look at the current impacts of CC in mainland China, in Japan, and Africa. Other negative impacts of CC include increases in diseases and death across several continents, a rise in crime rates, international prostitution networks, and child and adult slavery in factories throughout the planet. Yes, the *IWP international organization* will be monitoring and coordinating efforts across other international organizations.

Sure enough, **Scenario 4** in Chapter 8, *"Nuclear Warfare and Death of the Human Genome",* is one of the most apocalyptic scenarios because it addresses the event of a nuclear war in our planet. Among the main factors addressed in this scenario are a global total arsenal of 15,000 nuclear warheads in existence today, each one being 6,000 times stronger than the nuclear bomb dropped in Hiroshima, Japan, at the end of World War II, a probability of 1,000 Million deaths, "nuclear winter", electromagnetic warfare pulse (EMP), and radioactive fallout lasting for many decades. This chapter also reviews the more than 15 "nuclear exercises" carried over the last 65 years, the 10 "close calls" to having had a nuclear war, including the Cuban missile crisis of 1962, and the role of four major international organizations, including the North Atlantic Treaty Organization (NATO), and the Warsaw Pact. This chapter also explores potential coordinating activities of the *International World Organization for Peace (IWP)* among the nations with nuclear warhead capabilities.

Next, we address the following 4 scenarios in the category of *"negotiate-and-live"*, beginning with *Scenario 5, in Chapter 9, "Diversity of Religions and Beliefs"*, with a list of main factors, including a brief description of Christianity, Judaism, Islam, Buddhism, Hinduism, and Atheism. Other events visited in this chapter are Islam's golden Age and the Arab-Israeli conflict. Is it possible to get all these major religions to agree and constitute a *World Council of all Religions (WCR)* for purposes of extending the human life-cycle on Earth? The pillar and tenets of this council are listed and defined. Last, this chapter also considers the monitoring of conflicts, consulting, decision-making, and agreement making of the various religious hierarchies through the working of the triad WCR-IWP and United Nations (UN).

Is it possible to promote an economic and political balance of powers in the World? This is precisely the objective of *Scenario 6,* Chapter 10, *"Promote Economic and Political Balance across the Global Community"*. Among the main factors contributing to the contents of this chapter we find: current lack of political balance among nations today, lack of economic balance among many nations, vulnerable populations under the threshold of poverty, the ungoing "globalization" process, and major countries seeking rupture from traditional partnerships with other countries (e.g., *"Brexit"* in the United Kingdom). Very significantly, it outlines the 17 main goals of the *United Nations (UN)* international organization to bring about such a balance. It also explores the "good" and the "bad" attributes of globalization, and the creation of a new global order.

Did someone mention life in space colonies within our galaxy? That is exactly what we explore in **_Scenario 7_**, Chapter 11, **_"Search, Find, and Populate other Planets"_**. Main factors presented in this chapter include: the environmental degradation of Earth today, current space technology available today, the curiosity of our human species for unknown places, and the need to avoid more "close-calls" for nuclear wars on planet Earth. **_When could the first space colony become a reality_**? Very possibly in the year 2050 with initial ventures of colonization of the moon, our nearest planet. Many are the technological, social, and psychological factors to address, including ability of humans to co-exist in small space colonies over years and decades, ability and feasibility of humans to live in planet with much lower gravity forces (e.g., the Moon's gravity is only 1/6 that of planet Earth), the exchange of energy and minerals among space colonies, the high cost of launching human beings and materials to other planets, and the dynamics of parterships among Government and Private Sector corporations (e.g., joint ventures). A new world order would certainly emerge, impacting the role of citizen organization in the design and maintenance of space colonies, the frequency and magnitude of wars among nations on Earth, and the cost of solar energy for our towns and cities.

By now we have addressed technological events, social models, space colonies, overpopulation, and many other issues, but what if all of these events revolve around our own human nature. **_How do we describe and classify our own human nature_**? It is in **_Scenario 8_**, Chapter 12, **_"Coming to terms with our own Human Nature and Place in the Universe"_**, where we take a deep and serious look at our human nature developed over millions of years and kept within our genes today. Among the issues addressed in this chapter are: our origins as a species in Africa some 50,000 years ago, our unique place in the Universe, the absence of other forms of life in other planets and galaxies, the non-existence of a heaven, a hell, or afterlife and, most important, the recognition as individuals within our own social hierarchies. ***Is it possible to change our traditional ways of thinking about ourselves and the Universe, those beliefs which we have nurtured for thousands and millions of years***? Maybe yes, maybe no. But in this chapter we present some of the latest thinking in an effort to come to terms with this new knowledge and, if possible, attempt to extend our human life-cycle for thousands of years into the future.

Acknowledgements

Many are the persons that have contributed their effort in order to make possible the contents of this book. In first place, I would like to thank *Aloña Altuna*, my life and work companion, for her patience with my work and for her encouragement to see this book written, published, and distributed.

My special gratitude to *Dr. Lucien Duckstein, Dr. Wayne Wymore, Dr. Ferenc Szidarovszky, and Dr. Istvan Bogardi*, my dear professors then and friends today at the *University of Arizona*, Tucson, Arizona, USA, who inspired me with their research work, their quest for knowledge, for an understanding among people, and with their curiosity about people, our planet, and the Universe.

Chapter 1: Introduction

*"The **end of the world** is on people's minds. We have the power to destroy or save ourselves, but the question is what do you do with that responsibility?"*

--Nicolas Cage
Read more at:
https://www.brainyquote.com/quotes/keywords/end_of_the_world.html

*"The probability of **apocalypse** soon cannot be realistically estimated, but it is surely too high for any sane person to contemplate with equanimity."*

--Noam Chomsky
Read more at:
https://www.brainyquote.com/quotes/quotes/n/noamchomsk635905.html?src=t_apocalypse

Introduction

The story of our human species, the story of humanity, is an incredible story which goes back millions of years to remote times and places, full of challenges, uncertainties, a thousand civilizations and cultures, plagues, brief shining periods, wisdom, stupidity, new technologies, with an insatiable thirst for knowledge and discovery, and a will to go on against all odds. Yes, it has taken millions of years for our human species to learn of its origins in a micro-biological world, its relationship to all animal and plant species, and its existence as a speck of life in the Universe. *Intelligence?* Yes, we have come to develop several types of intelligence and trades which have enabled us to survive droughts, wars, poverty, and internal wars in each corner of the planet Earth.

And then there is the *irony* of it all. We were able to evolve and develop along millions of years, while our intelligence was primitive, basic, and merely functional. Other species such as those of the dinosaurs, sharks, insects, and even cockroaches have been able to live and evolve during thousands and millions of years with small but functional intelligences. In contrast, our species of Homo Sapiens Sapiens has been able to develop its intelligence in the last few thousand years, in part due to its existence and proximity to sea food, and today, in the 21-century we are ready to kill each other in a thousand ways and vanish from the face of the planet Earth. *Why?* While we were primitive and only functionally intelligent we were able to continue living forever, for thousands and millions of years. Now that we claim to be "intelligent", we have managed to design a variety of society models, a taste for intolerance in the face of differences in those models, resorting to corruption, fraud, crime, and large fratricide wars every 50-70 years. Be functionally intelligent, only, and live for millions of years; gain intelligence and acquire diversity of society models and be ready to die in a global holocaust. Why?

Is there time left and "intelligence" to design and explore other evolution paths for our human species? Guided by this question I propose we explore a *dichotomy of scenarios*: (1) "fight-and-die" scenarios, and (2) "negotiate-and-live" scenarios. In order to carry out this assignment we are going to need to have premises and guidelines, I suggest.

Contents of this chapter:
- **Scenarios, their design**
- **Time frameworks**
- **The Players**
- **Human needs**
- **Prevalent conditions**
- **Chain of Events**
- **Long term human condition**
- **Lessons to learn**

Scenarios, their Design

Why scenarios? The design of a scenario implies a set of plausible initial conditions from which a number of events may occur. It is also a way of organizing and grouping a large set of plausible circumstances, in our case plausible circumstances about our society models in our planet Earth. To that effect, I have opted for the design of four (4) *"fight-and-die"* scenarios, and another four (4) *"negotiate-and-live"* scenarios, as shown below.

"Fight-and-die" scenarios:

Scenario 1: Slow Death by Overpopulation, Food Shortages, and International Conflicts

This scenario is basically a representation and extension of the current social, economic, political, and technological conditions in the world today. Among these conditions stand prominently overpopulation, food shortages, internal conflicts, international conflicts.

21

Scenario 2: Control of Energy Sources, Guerrilla Warfare across the Planet

A planet with a mosaic of society models characterized by dozens and hundreds of internal and international guerrilla activities. There is no interest in a "common good" or "regional good" and, instead, countries ally themselves with partners of the same thinking. Flows of emigrants and political refugees augment across continents, especially from Africa to Europe.

Scenario 3: Climate Change and Environmental Holocaust

The polar regions are melting, sea water levels are rising, and coastal cities (e.g., New York, Los Angeles, Hong Kong, Dubai, Rio de Janeiro, etc.) are losing large tracts of urban development. Increased energy use to provide for transportation and relocation needs are contaminating the air, rivers, and agricultural lands.

Scenario 4: Nuclear Warfare and death of the Human Genome

The internal and international guerrillas have now turned into nuclear attacks among major powers. The first nuclear attacks were committed by small countries led by dictators (e.g., North Korea, Cuba, other), but soon the larger powers came into these international conflicts. Cities lie in ruins, vast agricultural areas have been contaminated with radiation. Flows of refugees all over the planet. Criminal groups rule the country side.

"*Negotiate-and-live* scenarios":

Scenario 5: Diversity of Religions and Beliefs

Religions are finally seen as power organizations and movements which often bring societies and countries into war. By now citizens across many countries unite to promote diversity among religious organizations in an effort to work together. Global education programs are initiated to share knowledge about the common origins of mankind. Citizens take over their societies, placing politicians on a secondary plane. All animals and plants are now recognized as members of the "family of species". Population control programs are established and monitored worldwide.

Scenario 6: *Promote Economic and Political Balance across the Global Community*

The world community is no longer waiting for war conflicts to erupt and then intervene. Instead, every country and society is monitored and assisted to attain minimum economic, education, and social standards. The intent is to prevent isolated areas which could be used by opportunistic political and criminal leaders to subdue their populations and used them in warlike activities.

Scenario 7: *Search, Find, and Populate other Planets*

Life in other planets in the Universe? Some individuals and communities believe that there is life to be found in other planets in the Universe, and this belief is used in this scenario to direct human energies to constructive programs and societies. All communities and countries united with a common goal. New space technologies are created and education programs are developed worldwide.

Scenario 8: *Coming to terms with our own Human Nature and Place in the Universe*

Finally, individuals and communities recognize the variety of conflicting factors housed in our *genes* over thousands and millions of years of evolution and development. The search is on for society models which bring in voluntary discipline to face these realities and to offer constructive relationships and means of co-existence. Hierarchies of power are recognized within each society, and individuals are encouraged to work at various levels in such hierarchies on a "rotational bases" in order to satisfy personal needs housed in our genes.

Time Frameworks

What are the time frames in which these scenarios will take place? One hundred years? Five hundred years? Ten thousand years possibly? There is really three time periods to consider: (1) the time period to the beginning of each scenario, (2) the duration of the scenario itself, and (3) the time period that follows which follows a scenario, either agonizing and dying as a species, or embarking into a new era of co-existence among the nations and communities on the planet Earth.

It is the belief of this author that the time left to the beginning of each one of the first 4 scenarios, the *"fight-and-die"* scenarios is in the order of a few decades only, possibly 100-200 years, no more, into the 21 century. We realize, hopefully, that a decade in the 21 century is the equivalent of 200-300 years of existence at the beginning of the 10th century, say, in terms of the speed and volume of events given our current population of 7,000 Million people.

What about the length in years of the scenarios themselves? We may be talking about 500 years, 1,000 years at most, given the massive destruction capability of our new technologies, and the speed of the emerging climatic changes.

So what happens after these fight-and-die scenarios approach their end? Our human species will not disappear from the face of the Earth, if that is what we are asking with this question. Instead, our human species would be decimated to wonder and linger over a toxic landscape during decades, possibly another one thousand years, organized in tribal groups dedicated to theft and pillage alongside existing roads and city neighborhoods. Criminal bands will march and search for food supplies which may remain stored in warehouses, killing other human beings in the process, as tribal members gather physical strength as their genetic makeup quickly deteriorates and dies along the way. Yes, not a pretty picture.

And what about the other 4 scenarios, the *"negotiate-and-live"* scenarios"? The time periods for these scenarios will be different altogether, for sure. If we consider *Scenario 5*, where the religions are abolished by mutual consent, the time period to the beginning of such scenario may be in the order of 500-1,000 years from the present, long after the eruption of many more war conflicts to come; it would be the communities of citizens themselves the ones to want such abolishment, given that the religious hierarchies themselves would want to perpetuate their status, mandate, and supremacy. What to do with the vacuum left after the abolishment of the vast multitude of religions in the planet? That would be a major problem. That empty space would have to be filled with other *aspirations* which satisfy our minds and bodies while we are still alive. It's pretty hard to replace the belief of a "heaven", an after-life full of rewards, happiness, and contact with the "gods", with earth-bound rewards. And then, again, perhaps the religions could never be replaced with any sequence of earth-bound rewards and, instead, we are encased on a path of self-destruction, all along maintaining that array of religions.

However, we could assume momentarily that the religions are abolished, and that the confrontation between communities due to economic, social, racial, and political differences are greatly diminished. What is then to be negotiated among communities and nations, and how is the *desire for co-existence and a long existence as a human species in the Universe to be sought*? That is another major problem which we will investigate in the following chapters, of course.

The Players

On each one of these scenarios many individuals, communities, and hierarchical powers would be participating in decision-making processes which would lead to many of the events already mentioned. Who, then, will be those players participating in the decisions to be made? Will the traditional powers represented by political parties, religious power hierarchies, dictators, and government officials constitute the decision makers, the players in those scenarios. I would hope that is not the case, otherwise we would continue the current path of search for power for a few and eventual self-destruction in those "*fight-and-die*" scenarios, in my opinion.

The citizens. The main players being *Kathy Thompson, Xabier Elurmendi, and Dr. Eugene Finley of the International*

World Organization for Peace (IWP). The communities of citizens. The people. They would have to be the new decision makers, I dare to say, in those four ***"negotiate-and-live"*** scenarios. But that would represent a major change in our way of carrying and managing events in our society? Yes, indeed.

Currently, ***in the 21 Century***, there is already a mosaic of decision-making models, some of which may yield to still larger changes in decision making. In ***Europe***, for example, individuals in the various political parties of each country determine the well-being and fate of its citizens; the ***political parties constitute the power and have the resources needed*** to enable changes in the social, economic, and political structures of each country; and, yes, the ***hierarchical religious powers*** also are the owners of vast resources of power through their influence on the minds of many citizens.

By contrast, citizens in communities in the ***Anglo-Saxon world*** have a larger control on their lives, compared to other fellow citizens in Europe, I would say. In most cities and towns in the USA, for example, there are several citizen organizations which look after a variety of interests, including cultural life, protection and care of animals, political postures, environmental concerns, and more. Also, the constitution and mandates of City Halls in these towns and cities enable people to collect signatures and remove politicians from office if the citizens so decide. Since when and how? Well, when our European relatives left Europe in the 16th, 17th, and 18th centuries on their way to the ***"Americas"*** they wanted to get away from their omnipresent governments and religious organizations in the "old country", in Europe. So, upon arrival in North America they made sure the town people would be able to direct their well-being and economies in the new towns and cities. What about mainland China, and other Asian countries? Yes, in that continent citizens are still subjects of the political parties and religious organizations; a major challenge to future generations who would like a more active and participative citizenship.

This is another major challenge awaiting future generations: the replacement of political parties and hierarchical religions organizations by citizen organizations. Citizen organizations first, political and religious organizations second.

A Model of Human Behavior, an Engineering Perspective [Return]

And how do we take into account the needs which we feel as human beings in this *set of scenarios*? The individual is born with a set of *selfish* instincts and needs, it could be as simple as that, I propose. The set of instincts are products of evolutionary processes over millions of years, and the needs are basic needs such as heat (energy needed for biological processes), food, sex pleasure, and social interaction, among others, as depicted in *Figure 1*.

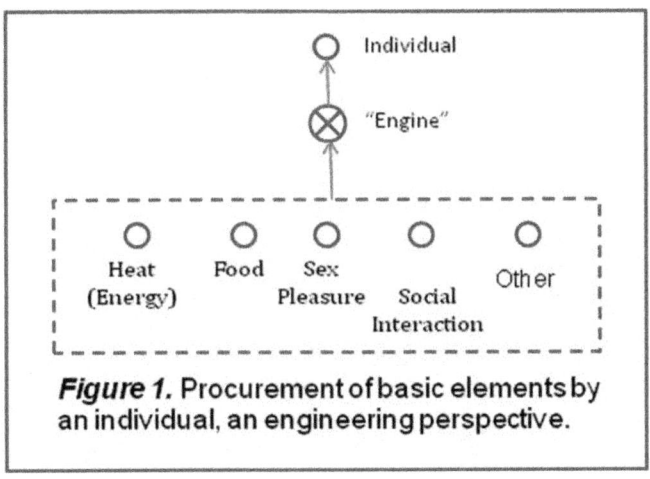

Figure 1. Procurement of basic elements by an individual, an engineering perspective.

Shown on *Figure 1* are three main components: (1) a set of basic elements (e.g., heat, food, sex pleasure, social interaction, other), (2) a process "engine", and (3) the individual (i.e., man, or an individual from any other species). The individual is trying to get to that set of basic elements to satisfy his/her basic needs, and *a process "engine"* is in the middle, between the set of basic elements and the individual, to enable the individual to gain access to those basic elements. What activities, if any, could be inside such a process engine? We might ask, and how do human abilities to satisfy basic needs, sets of values, religion (or lack of it), and other factors enter the picture? Good question, and to that effect I propose the contents shown in *Figure 2* below.

Now, within *Figure 2*, we can observe four other components: (1) a set of *means* to acquire basic elements, (2) a set of *objectives*, (3) a set of *strategies*, and (4) a set of *instincts*.

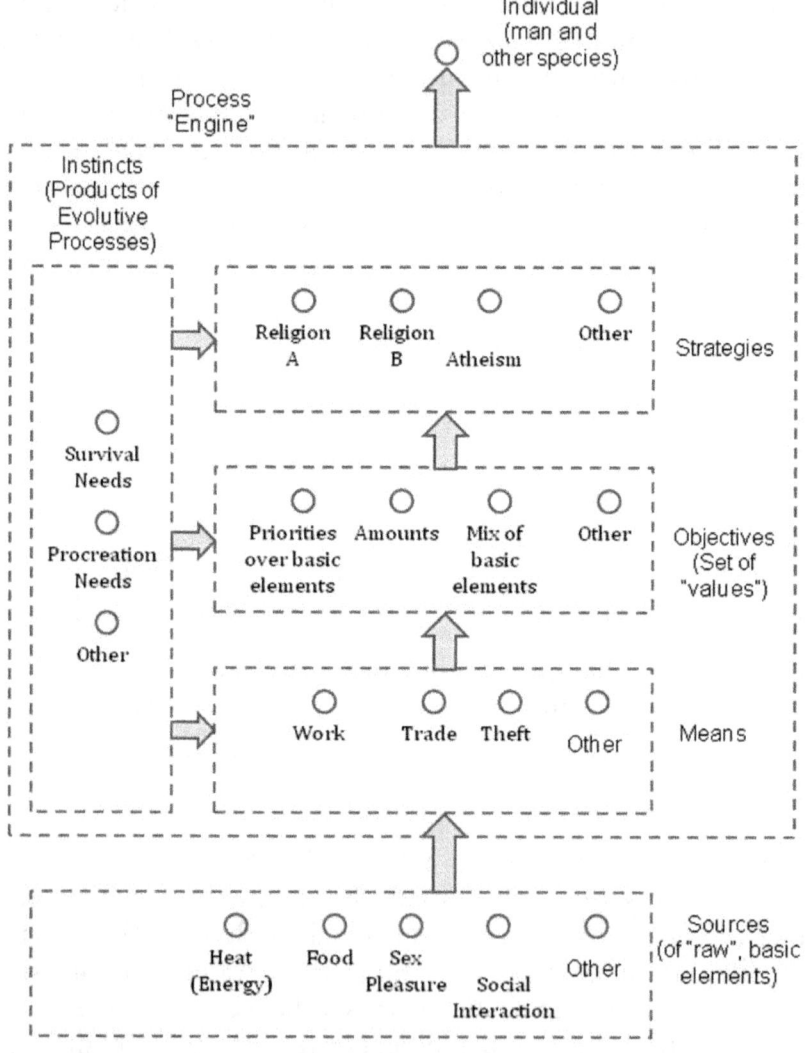

Figure 2. Procurement of basic elements by the individual, an engineering perspective of the evolutive process.

Set of basic elements. Heat (i.e., energy needed to carry out biological functions), food, sexual pleasures or activities, social interaction, respect, absence of fear, power, recognition, and others can part of this set of basic elements, some of which can be determined critical by an individual in order to carry out his existence.

Set of means. Various are the means that can be employed by an individual in order to acquire or try to acquire some of the basic elements, including activities such as working, trading, stealing, soliciting, and other. In the case of acquiring food, for example, the individual may resort to work in order to obtain wages with which to purchase food supplies; in the case of an individual with his unique set of "unlawful" values, such individual may opt for stealing some of the basic elements. Similarly, in the case of sexual favors, for example, the individual may opt for trading (e.g., food for sex), or other means (i.e., violence).

Set of objectives. These may include goals over the amounts of basic elements the individual wishes to acquire, the mix of basic elements, period of time of acquisition, and other. These objectives would also reflect the "***set of values***" of the individual regarding the relative value of one basic element over another (i.e., priorities). Yes, this set of values could include concepts such as "integrity", "morality", belief in a God or group of gods, belief in the absence of an after-life, "respect", "fairness", and other, as we have addressed in other chapters.

Set of strategies. Several are the strategies that *Homo Sapiens Sapiens* and individuals in other species utilize in order to achieve those basic elements, and these strategies can take many forms, frameworks, and ways of thinking, including Religion A, Religion B, Religion C, Doctrine A, Doctrine B, School of Thought A, School of Thought B, Sect A, Sect B, and even Atheistic Thinking A, Atheistic Thinking B, and so forth. Therefore, these strategies constitute ways available to the individual for purposes of just doing or rationalizing his/her actions in order to acquire those basic elements. For an individual within Religion A, for example, his strategy may be that of convincing others in his group that he has a special relationship with "God" or a group of Gods, an activity that requires all his efforts and concentration and, therefore, such strategy leaves no space for work of his own (e.g., as a farmer, a carpenter, a soldier, a brick layer, other), thus needing to be supplied with those basic elements by others in his society and religion. An individual in Sect A, as another example, may feel the need for sexual pleasures and activity from several women in his sect, and he/she may resort to convincing others in the sect that he/she is a prophet for "God" or group of Gods, and that copulation activity is seen as favorable in the eyes of such God. An Atheist individual, as another example, may feel the need for *social interaction* and discourse with members of his/her Atheist organization in order to learn about varieties of personal experiences in the absence of an after-life. And so forth.

Set of basic instincts. Included in this set are the individual's will to survive, and to procreate. Two very strong wants which are products of the evolutionary processes over millions of years. Together, these instincts exert powerful forces upon the individual's set of values, objectives, and strategies.

Prevalent Conditions

Each scenario is characterized by a set of prevalent conditions, the set of main set of social, economic, climatic, and political conditions that distinguish a scenario from others. **Scenario 1**, for example, is characterized by overpopulation and food shortages; **Scenario 2** is characterized by violent guerrilla warfare and large flows of refugees from one continent to another, while in **Scenario 3** climatic change plays a main role in the loss of large urban coastal areas, and so forth. These prevalent conditions will be defined in detail at the outset of each scenario in the following chapters.

Chain of Events

Each scenario will have its own chain of events, such that as many as 15-20 events would characterize each scenario. In **Scenario 4**, for example, a series of events would lead to the detonation of a first nuclear warhead by a country's army, causing thousands of deaths in another country and bringing in radiation conditions to dozens of towns and cities. Large migrations of peoples and animals would then take place from one region to other regions nearby. Other nuclear retaliation events would then follow, causing more people migration flows and the loss of vast tracts of agricultural lands, etc.

Final, Long-term Human Condition

The end of a time period for a *"fight-and-die"* scenario would not necessarily mean the end of the human species, not at all. Instead the human species would linger in the planet under drastic and painful conditions for hundreds of years, ranging from human settlements organized as "road-runners" armed with conventional arms, pillaging other communities for food and energy supplies, to urban settlements immersed in cannibalism and a steady deterioration of their human genomes due to the remnants of nuclear radiation in their surrounding lands, lakes, and forests nearby.

Similarly, the end of the time periods for the *"negotiate-and-live"* scenarios would produce major changes in the structuring of society and community models in the planet leading, hopefully, to frameworks of co-existence among nations and communities in the planet lasting thousands of years, compared to periods of hundreds of years only in the aftermath of the *fight-and-die* scenarios.

The Narrators

These scenarios are narrated by **Xabier Elurmendi**, **Kathy Thompson**, and **Dr. Eugene Finley** members of the *International World Organization for Peace (IWP)*, an international organization based in Tucson, Arizona, USA. As such, these scenarios are plausible scenarios, lasting hundreds of years each, articulated and detailed in order to ascertain manpower, technological, and material resources to be needed by this organization in the following decades. To that effect, Xabier, Kathy, and many other individuals, leaders in their own communities, have met repeatedly in order to postulate plausible scenarios for humanity over the next few thousand years, the next 2-3 thousand years. A result of their meetings has been the design and postulation of the 4 "fight-and-die" scenarios, and the other 4 "negotiate-and-live" scenarios presented and addressed in this book.

Lessons to Learn

The leaders of the *International World Organization for Peace (IWP)* hope that after the design and "running" of these 8 scenarios, there will be an opportunity to collect **lessons learned**, surveys, as gathered by all the participants in the design and "running" of those scenarios. ***Let the story begin.***

***

Chapter 2:
The Global Community Today

*"After the 9/11 **apocalypse** happened in New York City, people, particularly New Yorkers, who breathed in the ash, or saw the results of that, have a tendency to keep seeing echoes and having flashbacks to it."*

--**Stephen King**
Read more at:
https://www.brainyquote.com/quotes/quotes/s/stephenkin432568.html?src=t_apocalypse –

*"I don't believe in a **biological apocalypse**, but I think there is stormy biological weather ahead as the **human population** continues to grow."*

--**Richard Preston**
Read more at:
https://www.brainyquote.com/quotes/quotes/r/richardpre725568.html?src=t_apocalypse

Introduction

Where are we today, at the beginning of the 21st Century? If we are going to design and describe eight future scenarios of our human species in the planet Earth, it would make sense to try to summarize where we are today as a community of nations, I would suggest. Not an easy matter.

Contents:
- **Our geo-political network**
- **Population, statistics**
- **Differences of Religions**
- **Differences among Races**
- **Economic Conditions**
- **A history of global and local wars.**

Our Geo-Political Network

How do we begin to describe the planet Earth in which we live today? An impossible task? Seven continents, 174 countries, 3,036 languages, and 7,200 Million inhabitants. To make this task a bit easier, let us begin with one continent at a time. Ordered from largest in size to smallest, they are Asia, Africa, North America, South America, Antarctica, Europe, and Australia.

Asia. The Earth's largest and most populous continent, located primarily in the eastern and northern hemispheres. It covers an area of 17.2 Million square miles (44.5 Million square kilometers), or about 30% of the Earth's total land; a total of 48 sovereign states and members of the *United Nations (UN)* organization, including Afghanistan (Islamic Republic), Armenia, Bangladesh, Cambodia, People's Republic of China, India, Indonesia, Iran (Islamic Republic), Israel, North Korea, South Korea,, Mongolia, Pakistan, Philippines, Russia, Saudi Arabia, Turkey, and Vietnam.

Africa. The world's second largest and most populous continent, covering an area of 11.7 Million square miles (30.3 square kilometers), with 1,200 Million people; surrounded by the Mediterranean sea to the north, both the Suez Canal and the Red Sea along the Sinai Peninsula, the Indian Ocean to the southeast, and the

Atlantic Ocean to the west. A total of 54 countries, speaking 3.000 native languages.

North America. A continent in the Northern Hemisphere, bordered to the north by the Arctic Ocean, to the east by the Atlantic Ocean, to the west and south by the Pacific Ocean, and to the southeast by South America, and the Caribbean Sea. It covers an area of 9.5 Million square miles (24.7 Million square kilometers), with 23 countries, speaking hundreds of languages, including English, French, and Spanish; among those countries: Canada, Costa Rica, Cuba, Dominican Republic, Guatemala, Mexico, Nicaragua, Panama, and the United States of America.

South America. A continent located in the western hemisphere, over an area of 6.9 Million square miles (17.8 Million square kilometers), a total of 12 countries speaking hundreds of languages, including Portuguese, Spanish, English, French, Dutch, Quechua, and many other native languages. Among its countries we find Argentina, Bolivia, Brazil, Chile, Colombia, Ecuador, Paraguay, Peru, Uruguay, and Venezuela.

Antarctica. It is the Earth's southernmost continent, over an area of 5.4 Million square miles (14 Million square kilometers), surrounded by the Southern ocean. It is governed by 50 countries which have signed the *Antarctic Treaty System*. This treaty prohibits military, mining activities, nuclear explosions and nuclear waste disposal, but it supports scientific research with over 4.000 scientists from many nations.

Europe. With an area of 3, 930 Thousand square miles, a population of 742.5 Thousand people, and with over 225 languages, including English, Spanish, French, German, Italian, and Russian.

Australia. The world's 6th largest country by total area, with neighboring countries such as Papua New Guinea, Indonesia, and East Timor, over an area of 3 Million square miles (7.7 Million square kilometers). A population of 24.4 Million people, according to the 2011 census.

Population, statistics

We have come a long way since those 300 families left north-east Africa some 50,000 years ago.[3] Today we have a global population of 7,200 Million people and still growing, as *Figure 1* shows; by the year 2030 we may reach a population of 9,000 Million, and by the year 2100 we may well have a population of 16,000 if current birth rates continue.

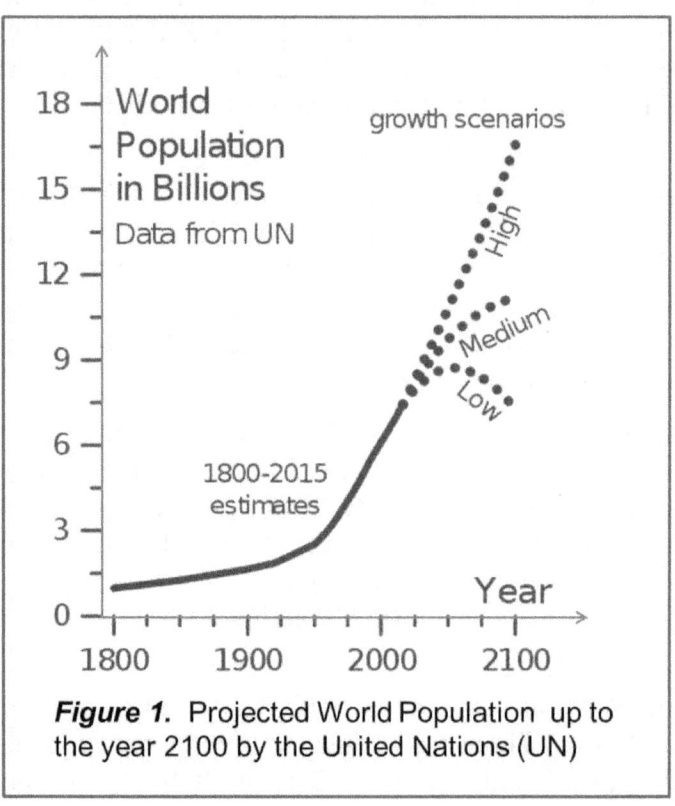

Figure 1. Projected World Population up to the year 2100 by the United Nations (UN)

Let us take a closer look with the help of *Figure 2*. In the case of Asia, for example, its population went from 1,500 Million to 4,500 Million in 2017, and it is expected to reach 6,500 Million in 2050. During the same time period, Africa has grown from 250 Million to 2,000 Million, and so forth. By combining the population growth across the six continents shown on that Figure 2, we may reach a total global population of 10,000 Million in the year 2050, in only another 33 years. Are we ready as a global community of countries and societies to receive such a future world with all the social, economic, religious, racial, and political problems which we have today?

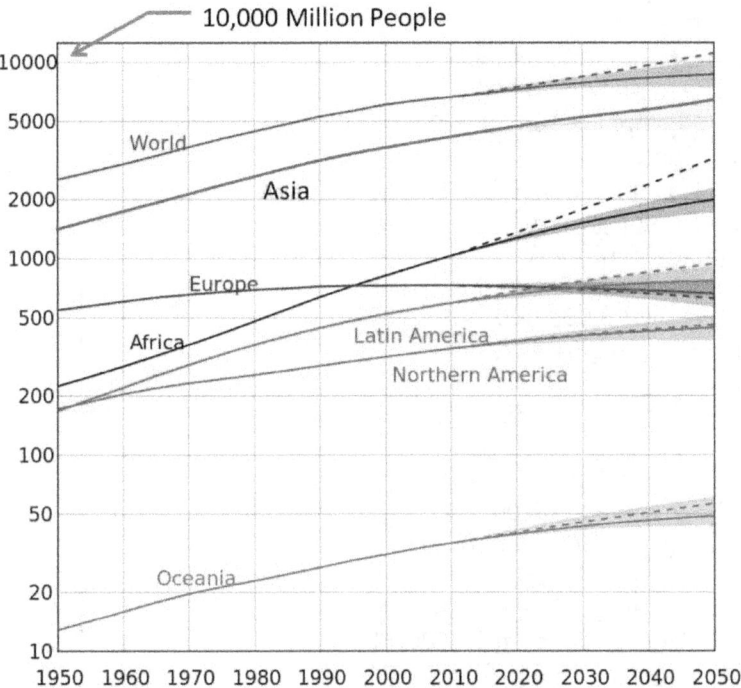

Figure 2. Projected Populations to the year 2050 by continents, in Millions. [1]

Differences of Religions

Our ignorance regarding the origin of our human species complicates drastically our understanding of the close relationship that exists among all races in the planet. Instead of understanding, of knowing that we have a common origin, that we are all family members, we complicate matters to the extreme with the myriad of religions all over the planet, each one giving its own explanation as to the origin of the Universe and the human species.

Let us begin by taking a look at the largest *Christian* populations in the world in *Figure 3*.

Christians
Largest Christian populations (as of 2011):
1. United States 229,157,250
2. Brazil 169,213,130
3. Mexico 106,204,560
4. Nigeria 80,510,000
5. Philippines 78,790,000
6. Russia 67,640,000
7. China 67,070,000
8. Democratic Republic of the Congo 63,150,000
9. France 55,948,600
10. Italy 55,832,000
11. Ethiopia 51,477,950
12. Germany 50,752,580
13. Colombia 44,502,000
14. Ukraine 41,973,000
15. South Africa 40,243,000
16. Spain 38,568,000
17. Poland 36,526,000
18. Kenya 33,625,790
19. Argentina 33,497,100
20. United Kingdom 33,200,417
21. Uganda 29,943,000
22. India 28,436,000
23. Venezuela 28,340,790
24. Peru 27,365,100
25. Indonesia 24,123,000

Figure 3. CHRISTIANS, largest populations by country [2]

By Christians we mean both Catholics and Protestants, of course. Notice that the population of Christians in these 25 countries add up to 2,200 Million believers, and that this number represents 31.50 % of the total global population of believers. United States leads in the list with 229,157,250 believers.

Next in the list of religions are the Muslims, as shown on *Figure 4*, with Indonesia leading the list of 20 countries with 206,,986,560 believers.

Muslims
A map of Muslim populations by numbers,
(Pew Research Center, 2009).
Largest Muslim populations (as of 2013):

1. Indonesia 206,986,560
2. Pakistan 180,608,292
3. India 172,000,000
4. Bangladesh 132,937,800
5. Nigeria 80,000,000
6. Iran 73,238,340
7. Egypt 70,056,000
8. Turkey 70,036,838
9. Algeria 36,092,810
10. Morocco 31,351,800
11. Afghanistan 30,112,680
12. Sudan 30,064,180
13. Iraq 29,767,300
14. Ethiopia 28,120,050
15. Saudi Arabia 26,624,560
16. Uzbekistan 25,628,240
17. Yemen 23,836,523
18. China 20,095,870
19. Syria 19,601,750
20. Malaysia 17,085,402

Figure 4. MUSLIMS, largest populations by country. [2]

This list of *Islam* believers in those 20 countries add up to 1,600 Million, which represents 22.32 % of the entire population of believers in the world.

Next is the world of *Buddhism*, with 10 countries represented in *Figure 5*. Notice that mainland China leads the list with 244,130,000 believers which add up to only 18.2 % of the total population in China, but which represents 50.1 % of the total global population of believers of Buddhism.

Countries with the largest Buddhist populations as of 2010			
Country	Estimated Buddhist population	% of the total population of the country	% of world Buddhist population
China	244,130,000	18.2%	50.1%
Thailand	64,420,000	93.2%	13.2%
Japan	45,820,000	36.2%	9.4%
Burma	38,410,000	80.1%	7.9%
Sri Lanka	14,450,000	69.3%	3%
Vietnam	14,380,000	16.4%	2.9%
Cambodia	13,690,000	96.9%	2.8%
South Korea	11,050,000	22.9%	2.3%
India	9,250,000	0.8%	1.9%
Malaysia	5,010,000	17.7%	1%
Subtotal for the ten countries	460,620,000	15.3%	94.5%
Subtotal for the rest of the world	26,920,000	0.7%	5.5%
World total	487,540,000	7.1%	100%

Figure 5. BUDDHISTS, largest populations by country. [2]

Next in the list of largest religions are the **Hindu** believers, as shown on **Figure 6.** As was to be expected, India has the largest population of Hindu believers with 957,636,314, followed by Nepal and Bangladesh.

Largest Hindu populations (as of 2010):
1. India 957,636,314
2. Nepal 21,354,570
3. Bangladesh 14,274,430
4. Indonesia 4,012,470
5. Pakistan 2,603,895
6. Sri Lanka 2,554,606
7. Malaysia 1,700,100
8. United States 1,543,730
9. United Arab Emirates 1,239,610
10. South Africa 749,870
11. Mauritius 665,820
12. United Kingdom 630,000
13. Canada 497,960
14. Tanzania 403,570
15. Kuwait 328,440
16. Australia 275,500
17. Singapore 264,370
18. Fiji 261,097
19. Trinidad and Tobago 240,100
20. Myanmar 203,000
21. Bhutan 177,100
22. Germany 120,000

Figure 6. HINDU, largest populations by country. [2]

This list of 22 countries of Hindu believers add up to 1,000 Million and represent 13.95 % of the total global population of believers.

Interestingly enough, the United States leads the 20 countries with followers of *Judaism*, as represented in *Figure 7*, followed by Israel with 5,907,500 and France with 493,600 followers.

Largest Jewish populations (as of 2011):
1. United States 6,588,065
2. Israel 5,907,500
3. France 493,600
4. Canada 375,000
5. United Kingdom 291,000
6. Russia 194,000
7. Argentina 181,800
8. Germany 119,000
9. Australia 97,300
10. Brazil 95,300
11. Ukraine 70,200
12. South Africa 67,000
13. Hungary 48,200
14. Mexico 39,200
15. Belgium 30,000
16. Italy 28,200
17. Chile 18,500
18. Turkey 17,400
19. Uruguay 17,300
20. Belarus 12,000

Figure 7. JEWISH, largest populations by country. [2]

Adding up the populations of believers in those 20 countries we arrive at 14.0 Million, which represents 0.20 % of the total global number of believers of all religions.

Next, on *Figure 9*, we have represented 20 countries with the largest population of **Bahá'i** followers, again with India leading those 20 countries with 1,897,651 and United States with 512,864.

Largest Bahá'í populations (as of 2010) in countries with a national population ≥200,000:
1. India 1,897,651
2. United States 512,864
3. Kenya 422,782
4. Vietnam 388,802
5. Congo, Democratic Republic of the 282,916
6. Philippines 275,069
7. Iran 251,127
8. Zambia 241,112
9. South Africa 238,532
10. Bolivia 215,359
11. Tanzania 190,419
12. Venezuela 169,811
13. Uganda 95,098
14. Chad 94,499
15. Pakistan 87,259
16. Burma (Myanmar) 78,915
17. Colombia 70,504
18. Malaysia 67,549
19. Thailand 65,096
20. Papua New Guinea 59,898

Figure 9. BAHÁ'I, largest populations by country. [2]

The populations in those 20 countries of Bahá'I followers add up to 7.0 Million, which represents 0.10 % of the total global population of believers of all religions.

Last, we present 18 countries with followers of *Jainism*, as shown on *Figure 10*, with India, again, leading such list with 5,146,697, followed by the United States with 79,459 believers.

Jainism. as of 2005:
1. India 5,146,697
2. United States 79,459
3. Kenya 68,848
4. United Kingdom 16,869
5. Canada 12,101
6. Tanzania 9,002
7. Nepal 6,800
8. Uganda 2,663
9. Burma 2,398
10. Malaysia 2,052
11. South Africa 1,918
12. Fiji 1,573
13. Japan 1,535
14. Australia 1,449
15. Suriname 1,217
16. Réunion 981
17. Belgium 815
18. Yemen 229

Figure 10. JAINISM, largest populations by country. [2]

Where are the other religious and non-religious populations represented?, we may ask. Precisely. Those populations are shown on *Figure 11*, where followers of *Atheism* add up to 1,100 Million representing 15.35 % of the total global populations of believers and non-believers, a significantly high number, we may say.

Religion	Adherents	Percentage
Christianity	2.2 billion[3]	31.50%
Islam	1.6 billion[4]	22.32%
Secular[a]/Nonreligious[b]/Agnostic/Atheist	≤1.1 billion	15.35%
Hinduism	1 billion	13.95%
Chinese traditional religion[c]	394 million	5.50%
Buddhism	376 million	5.25%
Ethnic religions excluding some in separate categories	300 million	4.19%
African traditional religions	100 million	1.40%
Sikhism	23 million	0.32%
Spiritism	15 million	0.21%
Judaism	14 million	0.20%
Bahá'í	7.0 million	0.10%
Jainism	4.2 million	0.06%
Shinto	4.0 million	0.06%
Cao Dai	4.0 million	0.06%
Zoroastrianism	2.6 million	0.04%
Tenrikyo	2.0 million	0.02%
Neo-Paganism	1.0 million	0.01%
Unitarian Universalism	0.8 million	0.01%
Rastafarianism	0.6 million	0.01%
total	7167 million	100%

Figure 11. Major Religions, total populations and by percentage. [3]

Economic Conditions

The economic well-being of humans also varies greatly across countries in the world, as *Figure 12* shows. The United States leads the list of 20 countries with a Gross Domestic Product (GDP) of 17,943 Billions (10^9) of dollars, followed by the European Union (EU) with a GDP of 16,220 Billions of dollars.

Rank	Country	Value (USD$)	Peak Year
—	*World*	73,171	2015
1	United States	17,947	2015
—	*European Union*	16,220	2015
2	China	10,983	2015
3	Japan	4,123	2015
4	Germany	3,358	2015
5	United Kingdom	2,849	2015
6	France	2,422	2015
7	India	2,091	2015
8	Italy	1,816	2015
9	Brazil	1,773	2015
10	Canada	1,552	2015
11	South Korea	1,377	2015
12	Russia	1,325	2015
13	Australia	1,224	2015
14	Spain	1,200	2015
15	Mexico	1,144	2015
16	Indonesia	859	2015
17	Netherlands	738	2015
18	Turkey	734	2015
19	Iran	665	2013
20	Saudi Arabia	653	2015

Figure 12. List of the 25 largest economies by GDP (nominal) at their peak level of GDP in Billions US $

These numbers can also be seen in groups, as shown on *Figure 11*. The major economies of the "G7" countries, for example, produce as much as 35,542 Billions of GDP yearly, which account for 46% of the world's yearly income. The group of China, India, Indonesia, and Thailand, qualified as "emerging and developing Asia" reach the figure of 14,944 Billion dollars, which is 19.3% of the total world GDP.

And what about the other side of the coin? Where are the people on the *poverty side* of the picture? Shown on *Figure 13* are the percentages of people in the various continents who manage to live with less than $1.0 a day. In Sub-Saharan Africa, for example, as many as 48% of the population are subsisting with less than $1.0 a day, followed by 33% (in 2001) in South Asia, and so on. Altogether, 23% of the world population manage somehow to subsist with such low economy.

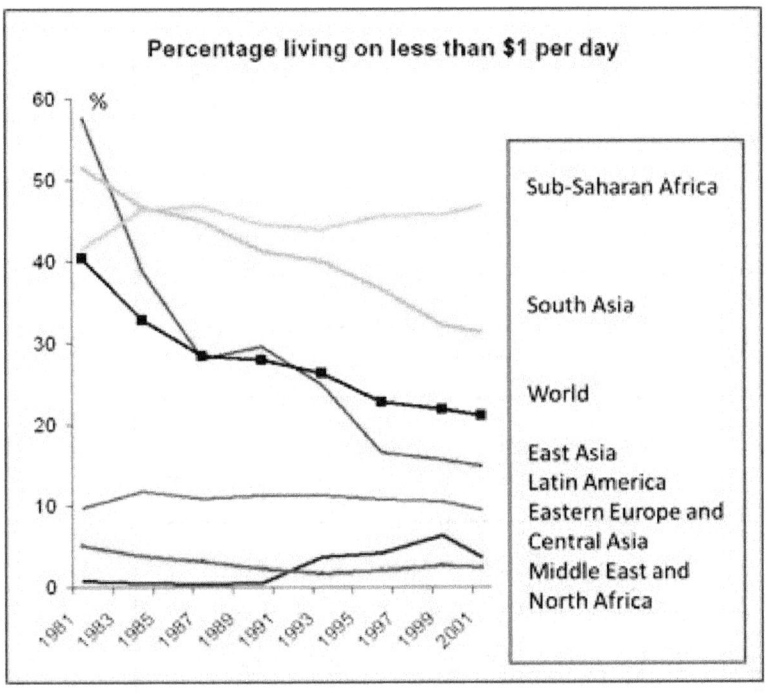

Figure 13. Percentage of the population by continent which live with less than $1 per day.**[5]**

Country Group	GDP (Nominal)	% of Global GDP	Number of Countries	Economies with at least 0.50% of Global GDP
Major advanced economies (G7)	35,542	46.0%	7	Canada France Germany Italy Japan United Kingdom United Stats
Emerging and developing Asia	14,944	19.3%	29	China India Indonesia Thailand
Other Advanced Economies (Excluding G7)	11,431	14.8%	30	Australia Austria Belgium South Korea Netherlands Norway Spain Sweden Switzerland Taiwann
Latin America and the Caribbean	5,799	7.5%	32	Argentina Brazil Mexico
Middle East, North Africa, Afghanistan, and Pakistan	3,458	4,50%	22	Iran Saudi Arabia United Arab Emirat
Commonwealth of States and Georgia	2,521	3.3%	12	Russia
Developing Europe	1,894	2.5%	12	Poland Turkey
Sub-Saharan Africa	1,68	2.2%	45	Nigeria
WORLD	77,269	100%	189	

Figure 14. List of country groups by GDP in 2014 (in Billions of US $). [4]

48

It is in *Figure 14* where we have grouped some of the countries with similar GDPs.

A history of global and local wars

With such enormous diversity of religions, economic conditions, social structures, and geo-political distributions, it should not be surprising to learn that the human species has been involved in hundreds and thousands of wars across thousands and millions of years.

Year	Battle	Country	Description
1854-1855	Siege of Sevastopol	United Kingdom	First use of the telegraph in combat.
1859	Austri-Sardinian War	France	First major use of railroads at the strategic level.
1861	First Battle of Bull Run	USA	First battle in which railroads play a decisive role.
1862	Battle of Hampton Roads	USA	First fight between two powered iron-covered warships.
1863	Battle of Gettysburg	USA	Largest battle ever fought in the Western Hemisphere.
1864-1865	Siege of Petersburg	USA	First case of modern trench warfare.
1866	Battle of Tuyuti	Argentina Brazil Uruguay Paraguay	Largest battle ever fought in South America.
1898	Spanish-American War	Spain USA	Extensive use of steel battleships in naval conflict.
1905	Battle of Tsushima	Russia Japan	Decisive battle between steel-covered warships.
1911-1912	Italo-Turkish War	Italy	First use of airplanes in combat.
1914	First Battle of the Marne	France	First large-scale of motorized infantry.
1914-1918	First Battle of the Atlantic	Germany	First campaign of submarine warfare.
1915	Second battle of Ypres	Germany	First large-scale use of chemical weapons in battle.

Figure 15. Countries involved in wars, and advances in war technology, in the time period 1854-1915

Year	Battle	Country	Description
1916	Battle of Verdun	France Germany	High point of fixed fortification warfare.
1917	battle of Cambrai	United Kingdom	First succesful use of massed tanks in combat.
1925	Rif War	France Spain	First modern amphibious assault using tanks and aircraft
1937	Bombing of Guernika	Germany Italy	First major use of terror bombing.
1940	Battle of Britain	Germany United Kingdom	First major war to be faught entirely in the air.
1940	Battle of Taranto	United Kingdom	First naval battle in which one side only employed aircraft.
1941	Battle of Crete	Germany	First major battle in which one side only employed airborn forces.
1941	Operation Barbarossa	Germany Soviet Union	High point of Blitzkrieg warfare, largest invasion in history.
1942	Battle of the Coral Se	Japan USA	First naval battle in which neither side's ships sighted of fired directly upon each other.
1942	Battle of Midway	Japan USA	Decisive battle between aircraft carriers.
1942	Battle of Stalingrad	Germany Soviet Union	Largest single battle in history, decisive battle of the Nazi-Soviet war.
1942	Battle of Guadalcanal	Japan USA	First major air-land-sea war in history.
1943	Battle of Kursk	Germany Soviet Union	Largest tank battle in history.
1944	Normandy Invasion	United Kingdom Canada	Largest seaborne invacion in history.
1944	Battle of Leyte Gulf	Japan USA	Largest naval battle in history.

Figure 16. Countries involved in wars, and advances in war technology, in the time period 1916-1944

Shown on *Figure 15* are the wars carried out in our planet for the time period 1854-1915, and on *Figure 16* for the time period 1916-1944.

Yes, a planet of primates continuously involved in fratricide wars, national and international wars.

Chapter 3:
Formulation of Criteria for the Eight Scenarios

*"I refuse to accept the view that mankind is so tragically bound to the **starless midnight of racism and war** that the bright daybreak of peace and brotherhood can never become a reality... I believe that unarmed truth and unconditional love will have the final word."*

--**Martin Luther King, Jr.**
Read more at:
https://www.brainyquote.com/quotes/topics/topic_war.html

*"I grew up in **war** and saw the United Nations help my country to recover and rebuild. That experience was a big part of what led me to pursue a career in public service. As Secretary-General, I am determined to see this organization deliver tangible, meaningful*

*results that advance **peace, development and human rights**.*"

--***Ban Ki-moon***
Read more at:
https://www.brainyquote.com/quotes/topics/topi
c_war.html

Introduction

In this chapter we would like to address two basic questions: (1) why ***eight scenarios*** and not some other number?, and (2) how does one go about ***designing those scenarios***? Exactly. The number of eight scenarios is completely arbitrary, I agree, although I would like this number to be sufficiently large to allow a number of diverse groupings of conditions among the community of nations in the planet; considering only two scenarios, for example, is too narrow to provide for the large variety of conditions which are currently developing and which may multiply in the near future. The eight scenarios, on the other hand, provide a "wide and open field" in which to postulate a number of "***starting set of conditions***" for each scenario. ***Criteria***, then, in the number of scenarios includes (a) variety of initial geo-political, social, economic, racial, and religious conditions which characterize our species in the planet, and (b) a range of conditions purposely wide enough to be able to address the complexity of the merging of those initial conditions and the large variety of environmental factors in our planet.

Next, what are the elements, conditions, factors to consider in ***the design*** of each scenario? We would like each scenario to be significantly different from the other seven scenarios, in terms of basic elements:

Contents:
- **Criteria:**
  - **Religions**
  - **Our primate nature**
  - **Population statistics**
  - **Food availability**
  - **Climate Change and Environmental factors**
  - **Inter-planetary Travel**

Criteria

How does one person, or a team of people, go about designing and later describing potential world scenarios which would predict the holocaust of the human species, or the beginning of a new and long era of understanding and well-being? Not an easy task. If the main idea is to generate *a set of world scenarios* so it includes the *one scenario* which would best represent the global outcome in the next 500-1,000 years, which are the elements and circumstances to consider as one describes the *set of events* in such scenario? Here are some thoughts that cross my mind:

Religions

If *religions* in primordial times, thousands and millions of years ago, provided our earliest societies the opportunity to organize and impose a set of co-existence rules, are they also responsible in the last 10,000 years for hundreds of wars among societies and communities with different views about *"Gods"* and *"the Universe"*? I can easily accept that the imposition of religions may have contributed to positive rules of "order" in those early societies, and even in today's global societies, no problem there. At the same time, we know, that those religions were used then and are used today by some *military leaders* "to capture minds and bodies" to send people to kill and die in wars. Also, people do no wait for children to grow up and to achieve reasoning abilities and world experience before they impose their religion on those children; consequently, those religious beliefs imposed during childhood are nearly impossible to eradicate or modify later on during adulthood. Religions, then, the large variety of religions which we already reviewed in Chapter 2, would need to be considered in the design of these world scenarios, one would say.

Our Primate Nature

There is much *irony* in our development and existence in the planet Earth as primates that we are, we, human beings. In ancestral times, particularly in the last 2-3 Million years, we have managed to survive other species through survival tactics which often included theft, killings of other species, murdering our own, and the imposition of force with the use of groups and tribes. If our ancestor primates run out of apples and other fruits in their own trees, they would then opt for raiding the trees and pastures of other groups of primates, stealing fruits, vegetables, and animals to use as food. If our ancestor primates had not opted for those strategies to survive and develop in the prairies, we would not be here today, I venture to say. We would have perished and vanished as a species. Period. We view today those strategies and events as "good" and "bad", but there are no such concepts when it comes to the survival of a species. I you want to live, then you need to eat food, wherever that food supply comes from, or you die.

Great. So our human species managed to survive and develop during thousands and millions of years. Built into our *genes*, then, were those strategies of theft, trade, killing of other species, etc., whatever was needed in order to survive. Today, in the 21st Century, we have those drives and strategies still built into our genes, and only our "society rules" are able to control those drives and strategies, most of the time.

Could it be, ultimately, that those drives and strategies which we developed and built into our genes, will eventually drive us to auto-destruction, fratricide, and death on planet Earth? *Quite an irony, indeed*. I'll say it again. Those traits and characteristics which helped us survive and develop in the very beginning, would also be responsible for our ultimate auto-destruction and death at the end of our life cycle as a species in the planet Earth. Unless, we come to terms with our very biological make up, our own way of thinking and behaving, our sometimes criminal nature, and opt for means to foster a long life cycle on planet Earth, that would be our fate, for sure, indeed. Possible?

Population Statistics

As we may recall from earlier chapters, our *Homo Sapiens Sapiens* species began with some 300 families who left Africa some 50,000 years ago, migrating into southern Europe, and later dispersing into all five continents. In that period of time we have managed to reproduce ourselves to reach *a population of some 7,000 Million*, in spite of several massive natural disasters due to erupting volcanoes, floods, and plagues as the ones that occurred in

Europe during the Middle Ages, a population that would otherwise be much larger today, at the beginning of the 21st century. How is a population of this size a threat to our species and the health of our planet? Good question. I have traveled all along the USA, several countries in Europe and Latin America, and I have seen many cities, yes, but also many vast areas of land with little population, being used mostly for agricultural purposes. Only a few years ago I travelled by car from the Basque Country south all the way to Madrid and, not taking the Burgos route, and I saw mostly empty spaces and agricultural lands, with very few small towns. In fact, much of the population in Spain and many other countries is to be found along its rivers and along its coastlines. So what is all this concern about overpopulation? Let us begin to look at the facts.

Pakistan, with 180 million people, 2.6% of the world's population, is the **sixth most populous country in the world**. Its total fertility rate (TFR) is 3 children per woman, and its economic growth rate is only 3%, compared with China's 9.2% and Bangladesh's 6.1%. Bangladesh has a TFR of 2.2. If Pakistan's TFR does not change, its population will reach nearly 380 million by 2050 and the country will face a devastating scarcity of resources, according to the UN. The Malthusian Theory of Population is still workable to understand the relation of population growth in **geometric means** and food growth in **arithmetic means** while technology remains constant. Unfortunately, the Government of Pakistan and political parties still do not have a clear vision to address this issue, imposing birth control panic with slogans like "**bachay do hi achay**" (Two Children are enough), which created further fear and confusion among masses.

Little or no progress has been made in **Baluchistan**, with a TFR of 4.1, Sindh and Khyber-Pakhtunkhwa 4.3, Punjab 3.9. Major political parties chant slogans for the empowerment of women but when it comes to women's health, they hesitate to include the population issue in their manifestos. The MQM is the only political party whose manifesto reflects the need for family welfare.[3]

We are still looking at Asia. So what is the overpopulation situation in the Philippines?

> The **Philippines**, a country the size of Arizona, has about 1/3 the U.S. population of 313 million and is expected to double in size by 2080. To feed its people, the Philippines imports more rice than any other country on the planet and its oceans show severe signs of overfishing. The Philippines has **one of the highest birth rates in the world** and the highest teen pregnancy rate in the Asian Pacific.
>
> Two thirds of native plant and animal species are endemic to the islands and nearly half of them are threatened. Less than 10% of the islands' original vegetation remains and 70% of the 27,000 square kilometers of coral reefs are in poor condition." Late last year Philippine President Benigno Aquino signed the *Responsible Parenthood and Reproductive Health Act of 2012*. This means that government health centers will have to make reproductive health education, maternal health care and contraceptives available to everyone. The **Catholic Church** is vehemently opposed to it and has threatened excommunication for the president and any politicians who support it. One 44-year-old woman, **a devout Catholic with 16 children**, said, "We don't pay attention to (the Church's opposition). They are not the ones who are giving birth again and again. We are the ones who have to find a way to care for the children." In the slums of its capital, Manila, a woman who had 22 pregnancies and has 17 surviving children, reported, "Many times, we sleep without eating." One of the reasons for enacting the reproductive health law is to help break the cycle of poverty. Pilot studies from USAID and UNFPA have shown that integrated population, health and environment (PHE) programs have made inroads in saving the environment.[5]

In some countries the fertility rate may be falling, but overall, the world population is exploding.

> There more than 3 billion people worldwide under the age of 25. About 1.2 billion of them are adolescents just

entering their reproductive years and there are political and cultural forces against contraception So even though birthrates are falling globally, the population explosion is far from over. In many parts of the world children are married at an early age, even at 10 or 11. Often they have babies as soon as they reach adolescence. If they choose, collectively, not to bend to parental and community pressure, and have smaller families than their elders did, the world's population -- now 7 billion -- will continue to grow, but more slowly.

According to UN projections, the number *will rise to 9.3 billion by 2050* -- the equivalent of adding another India and China to the world. This assumes that the worldwide average birthrate will decline from the current 2.5 children per woman to 2.1. If birthrates fail to fall, *population could reach 11 billion by midcentury* - the equivalent to adding three Chinas. Whether 9.3 billion or 11 billion, water, food and arable land will be more scarce, cities more crowded and hunger more widespread, but it will be worse with 11 billion.

John Bongaarts, a demographer at *Population Council in New York* told the Times, *"We're still adding more than 70 million people to the planet every year - which we have been doing since the 1970s."*

By 2030, India, now with 1.2 billion people, will probably see its birthrate drop from 2.5 children to 2.1. But even then, India's population will continue to grow because of momentum, and is not expected to peak until 2060, at 1.7 billion people. In some of the poorest parts of the world, fertility rates remain high, driven by tradition, religion, the inferior status of women and limited access to contraception. These are the same parts of the world where hunger, political instability and environmental degradation are already pervasive.

Africa is *expected to double in population by the middle of this century*, adding 1 billion people.

With 7 billion people in the world today, about 1 in 8 people lives in a slum and 1 billion are chronically hungry, according to FAO. At least 8 million die every year of hunger-related illnesses. When the population reaches 9 billion - around 2050 - 1 in 3 will be living in a slum, assuming poverty and migration to cities continue at their current rates. And there will be at least 2 billion more mouths to feed, but no one can say where the food will come from. *David Tilman*, a University of Minnesota expert on global agriculture, says **crop production will have to be doubled.** *William G. Lesher*, a former USDA chief economist, said **"We're going to have to produce more food in the next 40 years than we have in the last 10,000,"** he said. "Some people say we'll just add more land or more water. But we're not going to do much of either."

Most of Earth's best farmland has already been utilized, and cities and desserts are replacing it. Soil erosion, chemical contamination and salt buildup from irrigation are despoiling prime acreage. With climate change, higher temperatures and violent weather will stunt or destroy crops. Increased flooding will imperil millions living in low-lying regions.

But instead of worrying about this, in **Europe**, Japan and North America, leaders are worried about having too few young people to care for aging populations and to fund benefits for the elderly. And in developing countries, leaders often consider large youthful populations a source of economic vitality and political strength. In the U.S., political battles are being fought over contraception and abortion, causing some environmental and humanitarian groups to retreat from family planning initiatives.

Nearly 20 years after 179 nations signed a pledge to provide universal access to **family planning**, supplies of contraceptives remain erratic in much of the developing world.

Although *India*'s population growth has slowed among the urban middle class, birthrates remain high among the rural poor. *Uttar Pradesh*, a state in India 166 million people 10 years ago. Today it has 200 million and *may double by 2050*. If it were a country, it would be the fifth-most populous in the world. Women in the state still have 3.5 children each on average.

An extensive push is needed to make *contraceptives* widely available in scattered villages and rural areas, many of which lack paved roads or clinics. Government efforts have been haphazard and limited, reflecting ambivalence about family planning. A national law restricts women under 18 from marrying, but the tradition to marry early is still going strong. India's leaders view their country's youth bulge as a competitive advantage over China, whose workforce is older because of long-standing restrictions on family size. *Hania Zlotnik*, former director of the *U.N. Population Division* says: "But most of their growth is in the poor. Is it a good thing to have a larger number of poor people in your population?"

Advances in agriculture, followed by the Industrial Revolution, pushed humanity to the 1-billion mark around 1810. From there, the numbers began a steep ascent. *It took only 12 years to go from 6 to 7 billion.*

Although use of contraceptives worldwide has climbed steadily in the last 40 years, led by the industrialized West and China, it remains extraordinarily low in the least developed parts of Africa and South Asia.

In *Nigeria* only about 8% of reproductive-age women who are married or in relationships use contraception, compared with 72% in the US. Nigeria may surpass the U.S. as the third-most- populous country by 2050. *Kenya*'s family planning program was once held up as a model on the continent. In the late 1970s, the government joined with international donors in a high-profile effort that reduced the birthrate from more than eight per woman to fewer than five by the late 1990s.

Then **Kenya** was shaken by political turbulence, and a Republican-controlled U.S. Congress slashed family planning budgets. Supplies of contraceptives were interrupted across the East African nation and the decline in the birthrate stalled. The projection for Kenyans population in 2050 has been bumped up from 44 million to nearly a 100 million."[6]

There we have the **effects of overpopulation**: (1) an increase in poverty and disease among families worldwide, (2) abuse of women's rights, (3) depletion of natural resources, (4) degradation of the environment, and (5) conflicts and wars. Along this same issue, some of the **solutions to overpopulation** can be: (1) better education, (2) family planning, and (3) knowledge of sex education.

Food and land Availability

Vast amounts of foods are needed to maintain the world population today. In this section we take a look at the **changes** that the various types of foods, agricultural areas, and their statistics are undergoing due to a variety of reasons, including overpopulation and climate change.

The world needs to produce **at least 50% more food** to feed 9 billion people by 2050. But climate change could cut crop yields by more than 25%. The land, biodiversity, oceans, forests, and other forms of natural capital are being depleted at unprecedented rates. Unless we change how we grow our food and manage our natural capital, food security --especially for the world's poorest --will be at risk. Already, **high food prices** are the new normal. When faced with high food prices, many poor families cope by pulling their children out of school and eating cheaper, less nutritious food, which can have severe life-long effects on the social, physical, and mental well-being of millions of young people. Malnutrition contributes to infant, child, and maternal illness; decreased learning capacity; lower productivity, and higher mortality. One-third of all child deaths globally are attributed to under-nutrition. **Investment in agriculture and rural development** to boost food production and nutrition is a priority for the **World Bank Group**, which works through several partnerships to improve food security; from encouraging climate-smart farming techniques and restoring degraded farmland to breeding more resilient and nutritious crops to improving storage and supply chains for reducing food losses.[7]

In **Bangladesh** unplanned growth of population is complicating the process of meeting the **demand for food**, basic health requirements and educational facilities, which, in turn, is expected to lead to unemployment and social unrest. For example, trees are being chopped down for fuel on a regular basis. Climatic disruption in recent times, followed by salinity intrusion, shrinking farmlands and crop losses, has added to the woes of the people of the country. Bangladesh, with the world's highest density of population, is fast losing arable land due to growing industrialization and rapid encroachment of human habitat on farming areas. 8000 hectares of farm land are being lost every year from its original 13 million hectares of cropland due to urbanization, industrialization, unplanned rural housing and infrastructure buildings.

Entrepreneurs are going to the remote areas of the countryside to set up factories. Agriculture accounts for only 21% of the country's gross domestic product (GDP) although the sector employs around 50% cent of the nation's workforce. *At the current rate of loss of cultivable land, there will be none left in 50 years.* If the trend is not reversed now, the country would permanently lose its food security, making its poor population more vulnerable to volatile international commodity prices. The government has banned the use of arable land for purposes other than agriculture. It has been suggested that the factories and educational institutions that have already been built should now go vertical. But the government does not have adequate staff to monitor such things.

The average farm size has been reduced to less than 0.6 hectares and 59% of inhabitants are landless, with nearly

80% of the ultra-poor living in rural areas. 80% of Bangladesh's total cultivated area is in *rice*, the staple food and a politically-sensitive product. No one seems to worry about farmland depletion and the call for ensuring optimum utilization of arable land and bringing fallow land under cultivation is only rhetoric. Focus was put on rice production, while fuel, cooking oils and pulses were imported at volatile prices. Suggestions for *diversifying crops* have been ignored by the policymakers. Government expenditure on agricultural research has been steadily declining in Bangladesh. Investing more on agricultural research is vital for Bangladesh since it is losing cropland quite fast.

The *World Trade Organization (WTO)* pointed out that in the world's poorest corners, including Bangladesh, land is getting divided through inheritance and farm sizes are getting smaller and smaller with the passing of every generation.

The probable loss of arable and residential lands through flooding would result in increase of internal and external environmental migration and strained relations between countries. Bangladeshis, on an average, spend 50% of their income on food.

In Bangladesh, the problem of economic development has so far been addressed mainly in isolation from the population issue. It is expected that the *National Population Council* will play its due role in controlling population while strict monitoring and vigilance of RAJUK and all city corporations are a must to stop unplanned development of towns and industries across the country. The nation cannot afford to lose agricultural land any further. [4]

Next, researchers already see a *collision course* between overpopulation and climate change, a collision that will have a major impact on food supplies in the world.

We don't worry much about food, especially when it is plentiful and cheap. Only when prices rise do we pay much attention to any potential problems with the food supply. In the United States food has been relatively cheap for decades—typically costing less than 10 percent of our income—so we often take it for granted.

Perhaps we shouldn't. Quite a few people these days aren't taking food for granted, and it's not just the *more than 800 million people worldwide* who don't have enough to eat, or the *more than 47 million in the U.S.* who need food assistance. Whether we'll have enough food at affordable prices has been a particular concern for many scientists and economists since price spikes in 2008 caused unrest in places such as Egypt, Bangladesh, and Haiti.

This week in Washington (22 May 2014), the *Chicago Council on Global Affairs*, which for years has been immersed in questions about food and its supply—where it comes from, what kind and how much we grow, how we use or waste it, and whether we will have enough in the years ahead—gathered to discuss solutions to what its members see as an *emerging food crisis.*

These researchers see a collision ahead: between a rising *world population* that wants to eat more high-quality food such as meat and dairy, and a *climate system* that is diminishing harvests in many areas. Storms, floods, heat waves, and droughts are occurring with increasing frequency, trimming some crop yields across the planet.

That's a problem, because we will need more harvests of the major grain crops—*rice, wheat, and corn*—in the decades ahead. Those crops are the basis of nearly all the food we eat—even meat, because we feed grains such as corn, wheat, and soy to meat animals.

How do we meet the challenge of dramatically rising food demand? Jonathan Foley, a University of Minnesota researcher, is among many seeking solutions. In the May issue of *National Geographic*, he outlines a five-step framework for *feeding 9 billion people by mid-century.* Foley's article opens an eight-month series in the magazine on the future of food.

Beside the linkage between climate change and food supplies, this week's Chicago Council meeting focuses on improving harvests in vulnerable regions such as sub-Saharan Africa and South Asia, where harvests are low or unreliable and the need for food solutions is escalating. Topics include "sustainable intensification" of agriculture, which means getting better yields without damaging the land and water; improving irrigation in Africa; and using field schools to teach improved farming techniques.

Climate change gets some attention too. The group is examining how agriculture—which is responsible for about one-third of all greenhouse gas emissions—can reduce its impact on climate through precision tillage and fertilizer use. Discussants also emphasized the need for accelerated crop research to breed more climate-resistant crops that can survive heat waves, droughts, floods, and saltwater.

The group issued a report today that calls for the U.S. to make *food security*—investments in research, education, technology, and data—*a top priority for the long-term*, especially in the face of documented climate impacts on harvests.

The *U.S. Agency for International Development* also announced an initiative to help improve the nutrition of mothers and children in developing nations. It's part of a broader program called Feed the Future that aims to help alleviate hunger by helping farmers grow more and better crops across the developing world.

USAID Administrator *Rajiv Shah* told attendees at the Chicago Council meeting Thursday that Feed the Future has lifted more than 12 million children out of poverty and improved the welfare of more than 7 million farmers.

The *United Nations* projected last year that *the world's population would reach 9.6 billion by 2050*, up from nearly 7.2 billion now. A large portion of the increased food demand in decades ahead is projected to result from rising appetites for meat; several pounds of grain are needed to grow each pound of meat.

We are entering uncharted territory. The March report by the *Intergovernmental Panel on Climate Change* highlighted the vulnerability of food supplies to rising temperatures and extreme weather in years ahead. Studies already show that yields are being damaged by rising levels of carbon dioxide, increasing temperatures, heat waves, and droughts. The 2012 Midwest U.S. *drought*, for example, damaged crop yields and drove prices up.

Growing enough food for a booming world population as the climate changes is cited often as being *among the greatest challenges to face humanity*. Yet, young people who can help meet this challenge are not easy to find. One report this week indicated that U.S. agriculture is facing a shortage of trained scientists. So not only must we grow more food, we must grow more people interested in growing more food.[8]

Do we know that already a large number or countries are buying and leasing large extensions of agricultural lands in other countries in order to grow food such as *Africa, Australia, Latin America, and Siberia*?

The world's population is soaring past 7 billion. Food prices keep spiking every few years. Freshwater supplies in plenty of areas are dwindling.

And so, in response, a slew of countries and investors — from *Chinese state corporations to Gulf sheiks to Wall Street firms* — have started buying up farmland overseas in an apparent attempt to acquire as much precious soil and water as possible. This phenomenon is known as "*land grabbing*," and it has been accelerating ever since the massive surge in grain prices back in 2007.

So how much land and water is actually being grabbed? Quite a lot, according to a big new study published in the "Proceedings of the *National Academies of Sciences*" this week. The authors find that somewhere between 0.7 percent and 1.75 percent of the world's agricultural land is being transferred to foreign investors from local landholders. That's **an area bigger than France and Germany combined.**

Big purchasers of foreign farmland include Britain, the United States, China, the United Arab Emirates, South Korea, South Africa, Israel, India and Egypt. They're mostly seeking out land in **Africa and Asia**, particularly in countries such as Congo, Sudan, Indonesia, Tanzania, Mozambique, Ethiopia and even Australia. Here is a map from the PNAS study (Proceedings of the National Academy of Sciences of the United States, in www.pnas.org) showing who's grabbing what from where. Red triangles indicate investors, green dots indicate land that's being snatched up. Note that some countries, like Russia and Brazil, are on both ends of the farmland trade:

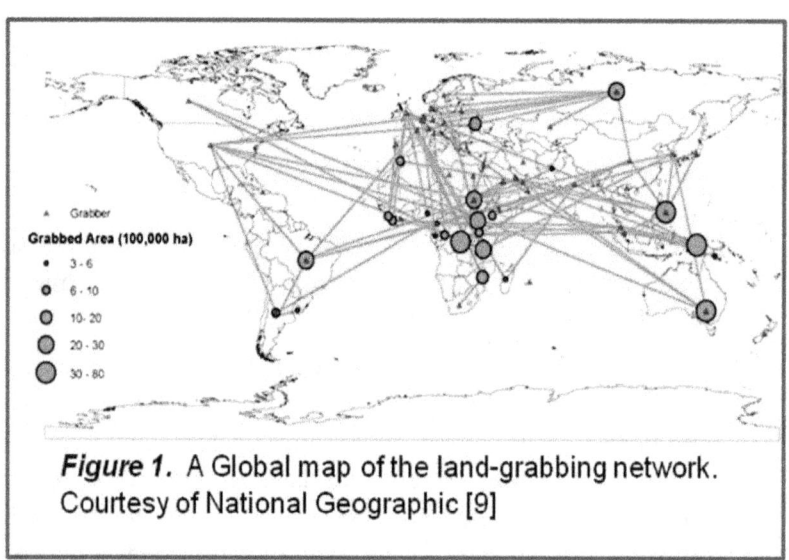

Figure 1. A Global map of the land-grabbing network. Courtesy of National Geographic [9]

The study found that foreign investors frequently buy tracts of land that have plenty of freshwater, either from local rainfall or underground aquifers. That's the key commodity here. "This is often good agricultural land that isn't yet fully utilized," says *Paolo D'Odorico*, one of the study's co-authors. "It was being used by local farmers without modern technology, without irrigation, without fertilizer."

After the land is bought up, large commercial farms will move in and boost production to grow their own crops. One 2010 study from the **World Bank** found that about *37 percent of this "grabbed" land is used to grow food crops, 21 percent to grow cash crops and 21 percent to grow biofuels*. (For instance, some 27,400 square miles of land have been snatched up in Indonesia, largely to grow palm oil, which can be turned into biodiesel.)

Is this a problem? In the abstract, it doesn't have to be — skilled foreign investors might be able to squeeze more use out of the land than the locals can. That could, in theory, be a mutually beneficial trade. But in practice, there are a lot of concerns about how all this land is actually being acquired.

Last year, for instance, **Human Rights Watch** released a report alleging that the Ethiopian government was forcibly relocating tens of thousands of people in order to lease land to foreign investors from China and the Gulf States. "The first round of forced relocations occurred at the worst possible time of year — the beginning of the harvest," the report said. "Government failure to provide food assistance for relocated people has caused endemic hunger and cases of starvation."

D'Odorico points out that a great deal of land and fresh water is also being bought up from poorer countries that are struggling to feed themselves, such as Tanzania. "If the food being produced on this land was going to locals instead of foreigners," he points out, "it would be possible for countries like Tanzania to cut down substantially on malnourishment."

Another example: The PNAS study notes that Sudan is leasing much of its prime farmland on the banks of the Blue Nile to foreign investors who are exporting food out of the country. Meanwhile, the rest of the people in this otherwise arid country have become increasingly dependent on food aid and subsidies.

There's still an enormous amount that researchers don't know about land-grabbing, according to D'Odorico. It took years of painstaking work just to assemble data on how much land and freshwater is actually being bought up abroad. And there are many basic things that researchers can't yet detail, including where the crops grown on grabbed land actually go, or how much yields improve when foreign investors come in.

What is likely, however, is that land grabbing will become more popular in the years ahead — especially if more governments start getting nervous about securing food supplies. *Fred Pearce*, who recently investigated the phenomenon for his book "The Land Grabbers," explained the upshot in an interview with the Guardian: "***The net result is that poor farmers and cattle herders across the world are being thrown off their land. Land grabbing is having more of an impact on the lives of poor people than climate change.***"[9]

There is still agricultural land out there available for grabbing, but the ***consequences are drastic*** in terms of massive displacement of people, failure to provide food to those displaced people by local governments or Private Sector, four-fold rise in rice prices, and capital-rich countries and organizations having an upper hand on food distribution.

The vanishing sea resources (we're eating all the fish!). A major problem and challenge here is that we are ***eating fish high up in the food scale***, such as salmon, tuna, shark, lobster, and many others. Very large and powerful vessel fleets are grabbing the fish without allowing time and space for their populations to grow and replenish,

Every year more than 170 billion pounds (77.9 million metric tons) of wild fish and shellfish are caught in the oceans—roughly three times the weight of every man, woman, and child in the United States. Fisheries managers call this overwhelming quantity of mass-hunted wildlife the world catch, and many maintain that this harvest has been relatively stable over the past decade. But an ongoing study conducted by *Daniel Pauly*, a fisheries scientist at the **University of British Columbia**, in conjunction with *Enric Sala*, a **National Geographic** fellow, suggests that the world catch is neither stable nor fairly divided among the nations of the world. In the study, called SeafoodPrint and supported by the Pew Charitable Trusts and National Geographic, the researchers point the way to what they believe must be done to save the seas.

They hope the study will start by correcting a common misperception. The public imagines a nation's impact on the sea in terms of the raw tonnage of fish it catches. But that turns out to give a skewed picture of its real impact, or seafood print, on marine life. "The problem is, every fish is different," says Pauly. "A pound of tuna represents roughly a hundred times the footprint of a pound of sardines."

The reason for this discrepancy is that tuna are apex predators, meaning that they feed at the very top of the food chain. The largest tuna eat enormous amounts of fish, including intermediate-level predators like mackerel, which in turn feed on fish like anchovies, which prey on microscopic copepods. A large tuna must eat the equivalent of its body weight every ten days to stay alive, so a single thousand-pound tuna might need to eat as many as 15,000 smaller fish in a year. Such food chains are present throughout the world's ocean ecosystems, each with its own apex animal. Any large fish—a Pacific swordfish, an Atlantic mako shark, an Alaska king salmon, a Chilean sea bass—is likely to depend on several levels of a food chain.

To gain an accurate picture of how different nations have been using the resources of the sea, the SeafoodPrint researchers needed a way to compare all types of fish caught. They decided to do this by measuring the amount of "primary production"—those microscopic organisms at the bottom of the marine food web—required to make a

71

pound of a given type of fish. They found that a pound of bluefin tuna, for example, might require a thousand pounds or more of primary production.

Nations with money tend to buy a lot of fish, and a lot of the fish they buy are large apex predators like tuna. Japan catches less than five million metric tons of fish a year, a 29 percent drop from 1996 to 2006. But Japan consumes nine million metric tons a year, about 582 million metric tons in primary-production terms. Though the average Chinese consumer generally eats smaller fish than the average Japanese consumer does, China's massive population gives it the world's biggest seafood print, 694 million metric tons of primary production. The U.S., with both a large population and a tendency to eat apex fish, comes in third: 348.5 million metric tons of primary production. And the size of each of these nations' seafood prints is growing. What the study points to, Pauly argues, is that these quantities are not just extremely large but also fundamentally unsustainable.

Exactly how unsustainable can be seen in global analyses of seafood trade compiled by Wilf Swartz, an economist working on SeafoodPrint. Humanity's consumption of the ocean's primary production changed dramatically from the 1950s to the early 2000s. In the 1950s much less of the ocean was being fished to meet our needs. But as affluent nations increasingly demanded apex predators, they exceeded the primary-production capacities of their exclusive economic zones, which extend up to 200 nautical miles from their coasts. As a result, more and more of the world's oceans had to be fished to keep supplies constant or growing.

Areas outside of these zones are known in nautical parlance as the high seas. These vast territories, the last global commons on Earth, are technically owned by nobody and everybody. The catch from high-seas areas has risen to nearly ten times what it was in 1950, from 1.6 million metric tons to around 13 million metric tons. A large part of that catch is high-level, high-value tuna, with its huge seafood print.

Humanity's demand for seafood has now driven fishing fleets into every virgin fishing ground in the world. There are no new grounds left to exploit. But even this isn't enough. An unprecedented buildup of fishing capacity threatens to outstrip seafood supplies in all fishing grounds, old and new. A report by the World Bank and the Food and Agriculture Organization (FAO) of the United Nations recently concluded that the ocean doesn't have nearly enough fish left to support the current onslaught. Indeed, the report suggests that even if we had half as many boats, hooks, and nets as we do now, we would still end up catching too many fish. [10]

Climate Change and other Environmental Factors

Our high industrial productivity worldwide also has its costs. The various forms of fuel (e.g., oil, gas, electrical power, other) must burn with oxygen in order to produce motion (e.g., rotation, vertical/horizontal displacement of components) needed for our machines to operate and function, thus releasing vast quantities of CO_2 into the atmosphere. It is that CO_2 that gathers into layers high up in out atmosphere thus creating a *"green house" effect*, not allowing Earth heat to radiate back into space. Result: the temperature of our Earth increases. We will learn more on climate change and its consequences in the description of the scenarios in the following chapters.

Inter-planetary Travel

On 4 October 1957 the *Soviet Union* launched into a low Earth Orbit the first artificial Earth satellite, *Sputnik*, thus giving way to a gigantic "space race" between Russia and the USA. Nine years later, the American *Neil Armstrong* became the first person to make a space flight, and three years later, in 1966, "Armstrong's second and last spaceflight was as mission commander of the Apollo 11 Moon landing- It is in Chapter 11, *Scenario 7*, where we will address ongoing activities towards space colonies in detail.

***

Chapter 4:
Weapon Technologies and Energy Sources

*"A society that admits misery, a humanity that admits **war**, seems to me an inferior society and a debased humanity; it is a higher society and a more elevated humanity at which I am aiming - **a society without kings, humanity without barriers.**"*

--**Victor Hugo**
Read more at:
https://www.brainyquote.com/quotes/topics/topic_war.html

*"**Peace** is not an absence of war, it is a virtue, a state of mind, a disposition for benevolence, confidence, justice."*
--**Baruch Spinoza**
Read more at:
https://www.brainyquote.com/quotes/topics/topic_war.html

Introduction

Primates that we are, we have always found ways to steal from each other, to trade food for sex and other commodities, to kill each other with sticks and rocks. The *"weapon technologies"* were primitive, yes, but they did not stop us from injuring and killing other primates in search of our objectives and desires, out in the jungle, out in the high planes. So why bring the subject of "weapon technologies" into this chapter? Volume, speed, depth, and surface covered are some of the reasons. If in pre-historical times our ancestor primates could kills dozens and hundreds of other primates and animals during one encounter, today we aim to kill and destroy thousands and millions of other human beings with more sophisticated weapons, providing "justification" with our diverse ideologies, of course. We have managed to do this harm to ourselves in countless fratricide wars in the last 50,000 years, in one and all five continents, as we already reviewed in Chapter 3.

If in ancestral times we fought and killed each other in order to protect *food supplies* and *energy sources*, today we continue killing each other in order protect food supply networks at national and international levels, as well as *to secure oil supply sources* over vast land and sea areas.

Contents:
- **Nuclear weapon technology**
- **Chemical warfare**

Nuclear weapon technology

Which countries have nuclear weapons, and how many of these weapons in total are there in the world today? We will provide answers to these two questions in this section with reference to country names, dates, and whether these countries belong to the *Treaty of Non-Proliferation of Nuclear Weapons (NPT)* or not. We may be surprised, however, at learning about the large number of such nuclear weapons.

To begin with, let us take a look at *Figure 1* which shows the number of warheads stockpiled by the United States and Russia in the time period 1950-2010.

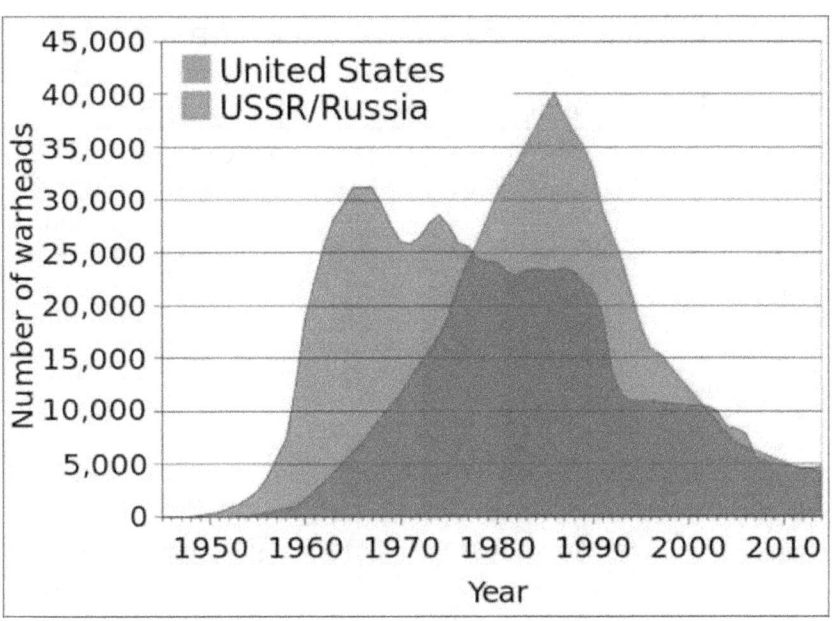

Figure 1. Number or Nuclear Warheads stockpiled by the United States and Russia over the time period 1950-2010. [2]

Sure enough, the number of warheads stockpiled by each one of these two countries is both staggering and alarming. Russia, for example, had managed to stockpile as many as 45,000 nuclear warheads in the year 1988, while the United States had 31,000 nuclear warheads in 1965. Do these numbers of warheads remain operational today, and who else has nuclear weapons?

There are 5 countries who are members of the ***NPT***, these being the United States, Russia, the United Kingdom, France, and China, as shown on ***Figure 2***.

Country	Number of Warheads (Active/Total)	Date of first test	CTBT Status	Delivery Method
Russia	1,790 /7,300	29 August 1949	Ratifier	Nuclear triad
United States	1,750 /6,970	16 July 1945	Signatory	Nuclear triad
United Kingdom	150 /215	3 October 1952	Ratifier	Sea-Based
France	290 /300	13 February 1960	Ratifier	Sea and air-based
China	n.a. /260	16 October 1964	Signatory	Nuclear triad
India	n.a. /120	18-may-74	Non-signatory	Nuclear triad
Pakistan	n.a. /130	28-may-98	Non-Signatory	land and air-based
North Korea	n.a. /10	9 October 2006	Non-Signatory	land and sea-based
Israel	n.a. /400	1960-1979	Signatory	Nuclear triad

Figure 2. Number of Warheads per Country-

Other countries included, but not belonging to the NPT, are India, Pakistan, and North Korea. *The United States* was the first country to develop nuclear warheads, carried its first test in July 16, 1945, and went on to use nuclear warheads to bomb and devastate the Japanese cities of Hiroshima and Nagasaki during World War II. Next, during the Cold War that followed, the US built some 70,000 nuclear warheads.

The *United Kingdom* (UK) went on to test its first nuclear weapon in 1952, working with refugee scientists who had left other countries in the European continent following the end of World War II. In fact, the UK, the United States, and Canada collaborated in the Manhattan Project in the development and testing of nuclear warheads. The UK deployed its nuclear weapons on bomber aircraft and on ballistic missile submarines (SSBNs).

By contrast, *France* conducted its own research and testing of nuclear technology on its own, motivated by the *Suez Canal Crisis* on the 1960s. After the Cold War France went on to disarm its nuclear capability by 175 warheads, reducing its arsenal, and deploying some 300 warheads of a dual system of submarine-launched ballistic missiles (SLBMs) and medium-range air-to-surface missiles, as noted on *Figure 1*.

Giant *China*, mainland, developed and first tested nuclear warheads in 1964 at the Lop Nur test site, as a deterrent against both the United States and Russia powers. It also tested its first "hydrogen bomb" in 1967, a very short time of 3 years after its first test. It is believed to have arsenal of 400 nuclear warheads today.

Among the countries which do not belong to the NPT is *India*, which went on to test and detonate its first nuclear warhead in 1974, calling it a "peaceful nuclear explosive." In 2005 US President *George W. Bush* and Indian Prime Minister *Manmohan Singh* announced plans to reach an indo-USA nuclear agreement. Today India still maintains an arsenal of 120 nuclear warheads.

Pakistan is not a member of the NPT organization either, but frequently it has been at odds with India. As such, Pakistan has been developing and testing nuclear warheads since the late 1970s with equipment and materials supplied by Western powers. It is estimated that this country maintains a stockpile of 130 nuclear warheads.

And then, there is *North Korea*, originally a member of the NPT treaty until it announced its withdrawal in 2003, after the USA accused it of having a secret uranium enrichment program. It conducted its first nuclear testing in 2006 claiming intimidation by the United States. Ever since, North Korea has continued in its development and testing of nuclear warheads, extending such activity to 2017, in the middle of a dramatic "Cold War" crisis with the United States.

Israel? We almost forgot about Israel! Yes, this country also has nuclear weapons, although it has not come forward to actually confirm so. *"Israel is not a party to the NPT, and purposely it engages in strategic ambiguity"*, is said in the global community. According to the Federation of American Scientists, Israel very likely has around 75-400 nuclear warheads for delivery in Jericho II medium-range ballistic missiles, of which 30 are gravity bombs to be delivered by aircraft.

More nuclear warheads out there in the planet? Yes. Under the **North Atlantic Treaty Organization (NATO)**, the **United States** shares some of its own nuclear arsenal with several countries, including Belgium, Germany, Italy, Netherlands, and Turkey, as shown on **Figure 3**. Both USA forces and local armed forces join to maintain this capability against the possibility of a Russian attack. Belgium, for example, stockpiles 10-20 nuclear warheads at its air base in Kleine Brogel under the management of the 52nd Fighter Wing.[2]

Country	Air Base	Custodian	Warheads
Belgium	Kleine Brogel	52nd Fighter Wing	10 --20
Germany	Büchel	52nd Fighter Wing	20
Italy	Ghedi Torre	52nd Fighter	10--20
	Aviano	31st Fighter Wing	50
Netherlands	Volkel	52nd Fighter Wing	22
Turkey	Incirlik	39th Air Base Wing	60--70

Figure 3. USA Nuclear Warheads in host Countires. [2]

Chemical warfare

We define chemical warfare as the activity which makes use of *toxic properties of chemical substances* as weapons. For how long have human beings been using chemical weapons in war conflicts? The answer is forever. Yes, it appears chemical weapons have been used by human beings since ancient times, as we propose to review next.

Ancient *Greek* mythology already mentions *Hercules* poisoning his arrows with venom from the Hydra monster in order to fight enemies. In the *"Laws of Manu"*, a *Hindu* book forbids the use of poison and fire arrows, although it advises poisoning food

81

and water. Arsenical smokes were known to the **Chinese** as early as 1,000 BC (Before Christ). Already in the second century BC writings of the Mohist sect in China describe the use of bellows to pump smoke from burning balls of mustard into the tunnels being dug by an enemy army.[1]

Similarly, there are numerous episodes and events in early modern era where the use of chemical weapons was employed. The historian *David Hume* describes how during the reign of *Henry III* in **England**, the English Navy destroyed an invading French fleet (1216-1272) by blinding the enemy sailors with "quicklime" the old name for *calcium oxide*. *Leonardo da Vinci* himself had proposed the use of powder of sulfide, arsenic, and verdigris in the 15th Century:

> *"Throw poison in the form of powder upon galleys. Chalf, fine sulfide of arsenic, and powdered verdigris may be thrown among enemy ships by means of small mangonels, and all those who, as they breathe, inhale the powder into their lungs will become asphyxiated."*

During sieges to cities in the 17th Century, armies tried to start fires by launching fire balls filled with **sulfur, tallow, rosin, and turpentine**; the resulting smoke and fumes provided considerable distraction among the city population.

During the *Industrial Era* in the mid-19th Century, the use of chemical weapons advanced. *Lyon Playfair*, Secretary of the Science and Art Department proposed the use of a *cadodyl Cyanide artillery shell* against enemy ships during the **Crimean War of 1854**; however, a British Ordnance rejected the proposal saying *"as bad a mode of warfare as poisoning the wells (water) of the enemy."* And what about the use of chemical weapons during the American Civil War? Well, the subject was considered; a New York school teacher by the name of John Doughty proposed the use of chlorine gas, delivered by filling a 10-inch artillery shell with 2 quarts of liquid chlorine which would produce many cubic feet of chlorine gas. This proposal, however, was never carried out.

Yes, there were agreements on the prohibition and use of chemical weapons by the 19th Century. The *Hague Declaration of 1899* and *the Hague Convention of 1907* forbade the use of poisoned weapons in warfare conflicts. And yet, some 124,000 tons of gas had been produced by the end of World War I and, in fact, the **French** were the first to use chemical weapons during that war using *tear gas ethyl bromoacetate.*

Shown on *Figure 4* is a list of chemical agents, as these have been used in wars during the time period 1914-2000. For each chemical agent, the list shows the method of dissemination of the agent, detection, and equipment used for protection.

Year	Agents	Dissemination	Protection	Detection
1914	Chlorine			
	Chloropicrin	Wind	Gas masks	Smell
	Mustard Gas	Dispersal		
1918	Lewisite	Chemical Shells	Gas masks	
1920s		Projectiles with	CC-2 clothing	
		Central Bursters		
		Aircraft		Blister
1930s		bombs		agent
				detector
		Missile Warheads	Protective	
1940s		Spray tanks	ointment	
			Gas masks	
			with Whetlerite	
1960s	V-Series	Aerodynamic	Gas Mask	Nerve
	nerve agents		with water supply	gas
				alarm
1990s	Novichok	Binary	Improved gas	Laser
	nerve agents	munitions	mask	detection

Figure 4. Chemical weapon technology. [1]

Approximately 70 different chemical warfare agents (CWA) have been used during the 20th and 21st Centuries in the form of liquid, gas, and solid form, as depicted in *Figure 5* and *Figure 6*. These agents are inert agents, and come in four categories: ***choking, blister, blood, and nerve***, depending on the system of the human body that they attack.

Highly important is the method of ***delivery*** of the chemical warfare agent, as we may suspect, and briefly describe next.

Dispersion. Placing the chemical agent adjacent to a target immediately before dissemination. A first use of this technique occurred during World War I, where the ***French forces*** used a 26 mm cartouche with tear-producing *ethyl bromoacetate* fired from a flare carbine.

Class of Agent	Agent Name	Mode of Action	Signs and Symptoms	Rate of Action
Nerve	Cyclosarin Sarin Soman (GD) Tabun (GA) VX VR Insecticides Novichok agents	Inactivates *enzime acetylchol* ineste-rase, preventing the breakdown of the neurotransmitter acetylcholine.	Miosis Blurred vision Headache Nausea, vomiting Sweating Muscle twitching Dyspnea Loss of consciousness.	Vapors: seconds to minutes Skin: 2-18 hours
Asphyxia nt/Blood	Most Arsines, Cyanogen chloride, Hydrogen cyanide	*Arsine:* Causes intra-vascular hemolysis and may lead to renal failure *Cyanogen chloride:* prevents cells from using oxygen.	Possible cyanosis Possible cherry-red skin, Nausea Patients gasp for air Seizures prior to death.	Immediate onset.
Vesicant/ Blister		Agents are acid-forming compounds which damage skin and respiratory system	Severe skin, eye, and mucosal pain. Skin erythema with large fluid blisters. Tearing, corneal damage. Mild respiratory distress.	*Mustards:* 4-6 hours eye and lungs affected. Skin: 2-48 hours *Lewisite:* immediate.

Figure 5. Classes of Chemical Weapon agents (Part 1 of 2). [1]

Thermal dissemination. It is the use of explosives to deliver chemical warfare agents (CWA). This technique was developed in the 1920s, and it allowed de delivery of significant amounts of an agent to be disseminated, spread, over a considerable distance. Not a very efficient device as much of the agent is initially lost by incineration.

Class of Agent	Agent Name	Mode of Action	Signs and Symptoms	Rate of Action
Choking/ Pulmo- nary	Chlorine. Hydrogen chloride. Nitrogen oxides. Phosgene.	Pronounced respiratory pains, flooding lungs and causing suffocation	Airway irritations. Eye and Skin irritation. Sore throat. Chest tightness. Wheezing. Bronchospasm.	Immediate to 3 hours.
Lachrima- tory agent	Tear gas. Pepper spray.	Causes severe stinging of eyes and temporary blindness.	Powerful eye irritation.	Immediate
Incapaci- tating.	Agent 15 (BZ)	Causes atropine- like inhibition of acetylcholine. Causes peripheral nervous system effects.	May appear as mass drug intoxication with erratic behavior. Hypothermia Ataxia (lack of coordination) Mydriasis (dilated pupils)	30 minutes to 20 hours. Skin: 36 hours Duration: 72-96 hours.
Cytotoxic proteins	Biological proteins: Risin, Abrin	Inhibits protein synthesis.	Latent period of 4-8 hours followed by flu-like symptoms. Nausea, cough Ingestion: Gastrointestinal hemorrage. eventual liver and kidney failure	4-24 hours:

Figure 6. Classes of Chemical Weapon agents (Part 2 of 2). [1]

Aerodynamic dissemination. A non-explosive technique to deliver a chemical agent from an aircraft, allowing the winds to disseminate the agent, and initiated in the 1960s. A more efficient technique as it is possible to model the altitude of dissemination, wind direction and velocity, as well as the direction and velocity of the aircraft carrying the agent.

Chapter 5:

Scenario 1: Slow death by Overpopulation, food Shortages, and Internal Conflict

*"The **human overpopulation issue** is the topic I see as the most vital to solve if our children and grandchildren are to have a good quality of life."*

-- Alexandra Paul
Read more at:
https://www.brainyquote.com/quotes/keywords/overpopulation.html

*"We already have the statistics for the future: the growth percentages of pollution, **overpopulation**, desertification. The future is already in place."*

-- Gunter Grass
Read more at:
https://www.brainyquote.com/quotes/keywords/overpopulation.html

Kathy Thompson

--Are we ready for the presentation of the eight scenarios to the members and invited guests of the *International Organization for World Peace (IWP)* this afternoon? Asked *Kathy Thompson* as she looked at her team mate *Xabier Elurmendi* across the conference table.

A few seconds of silence, as Xabier organized his thoughts.

--Yes, I think so, Kathy. After all, the two of us have been preparing for these scenarios during the last six months, as requested by *Dr. Finley*, our boss at the IWP. Of course, this afternoon we will be talking about *Scenario 1* only, on overpopulation and food shortages, as you may recall. It is a drastic scenario with lots of factors overlapping which may bring in a good number of questions from the audience, but we will be able to bring out our main points, I believe.

--Yes, you are right, it is a very complex scenario bringing the world to its very limits, but we need to address this scenario and the others if the IWP organization is to prepare participating countries for this outcome, really. How about going to the cafeteria to have a cup of coffee or tea before coming back to this conference room to meet members and guests?

Once at the cafeteria, both Kathy and Xabier review in their minds the contents of the 25-30 Powerpoint slides they have prepared for Scenario 1.

--Don't forget, Kathy, you go first with the initial 15 slides and then I will take over to cover the rest of the scenario, right? Ahh, yes, almost forgot, there is a short question-and-answer period at the beginning of the talk, the presentation follows, and then another final question-and-answer period at the end.

--Do we entertain questions from the audience during the presentation itself? Asked Kathy, holding her cup of tea with both hands.

--Yes, of course, no problem there. In fact, it all should help with the presentation and the recommendations that we have to prepare later, I would say.

Half an hour later Kathy and Xabier are back at the large conference room in the *Department of Systems and Industrial Engineering Department* of the **University of Arizona**, in Tucson, Arizona, USA. Some 45-50 people are already on their seats while another 4-6 people are still walking around the room.

--Good afternoon, Ladies and Gentlemen, welcome to our IWP series, today being the time to go over our *Scenario 1* with title:

"Slow death by Overpopulation, food Shortages, and Internal Conflict"

My name is **Kathy Thompson**, from the *Systems and Industrial Engineering Department*, as well as a member of *the International Organization for World Peace (IWP)* under the direction of **Dr. Eugene Finley**, whom most of you already know. I will present the first 15-20 powerpoint slides and will try my best to answer your questions along the way. Next, my partner **Xabier Elurmendi** will follow with the rest of the presentation, and he will also try to answer your questions during that second half. Any questions to get started?

Number of Scenarios

An arm is raised in the audience.

--How many scenarios has your team put together so far, what are the objectives of creating these scenarios, and what criteria did you all follow to create these scenarios? I mean, are they realistic?

--Thank you for your questions, much appreciated. Well, our team at the University of Arizona, here in Tucson, has been asked by the *International Organization for World Peace (IWP)* to design and create a number of "extreme scenarios" of living conditions, war conflicts, and potential agreements among conflicting countries and interests. A main idea is to be able to come up with and identify a set of **extreme conditions** in the planet over the next 200-300 years and, in the process, prepare a list of recommendations to the IWP and participating countries so that we are able to anticipate and prepare for such events. Said differently, many individuals and countries who are members of IWP believe that our human society is risking auto-destruction, the way we are going about solving our problems with overpopulation, shortage of food resources, religious wars, and shortage of energy resources.

--Yes, I understand, but why **eight** "extreme scenarios"? Why not 4-6 scenarios?

--You are right, the set of 8 scenarios is completely arbitrary. We did begin with the idea of creating 4 scenarios, but as we looked into the many complexities involved slowly we grew to 8 scenarios. In fact, those first 4 scenarios we called "**fight-and-die**" scenarios, but after a few weeks we realized that we also needed to address another set of "**negotiate-and-live**" scenarios in order to address both sides of the extreme bar. So today we are going to share with you the contents of Scenario 1 in that first set of "fight-and-die" scenarios, OK?

--I like that, thank you!

--You are very welcome, madam. On this first PowerPoint slide you see a list of what we call "main factors", that is, the conditions which characterize this Scenario 1:

- **Overpopulation**
- **Food shortages**
- **Countries buying land in Africa and in New Zeland**
- **Criminal Road and Sea Mafias**
- **Crime statistics rising, and**
- **Collapse of World order**

Overpopulation, statistics by country

In terms of *overpopulation*, to get started, we have already reached the number of *7,200 Million people in our planet Earth*, and that is a staggering number if we consider that it was only some 300 families that left mother Africa 50,000-70,000 years ago, on their way to Europe, the Middle East, and the other continents. How did we, the *Homo Sapiens Sapiens*, get to reach such a high number in population with all the plagues and wars which we have had over those thousands of years? Well, if you apply some basic mathematics, and some basic conditions per family, you would end up with such high population today. Just try it. It gets worse, really, as the population in our planet is expected to reach the 9,500 Million people in the year 2050, only a few years away into the future.

Time Period

Another hand is raised in the audience.

--What is the time period that we are talking about? Is this scenario to occur in 5,000 years, 10,000 years, or when?

--Thank you, sir. Wish we had that much time left as a species, but that is not the case. We are taking about this Scenario 1 happening over a period of 200-250 years beginning just now. Yes, conditions and matters are getting so extreme and drastic that such scenario is about to begin.

Sequence of events

--In previous presentations we have already shared with you the statistics on *food shortages* all over the world, as well. We have brought to your attention how countries like *China, Russia, Germany, the United States, Japan*, and another 15-20 countries have been leasing lands in Africa and in New Zealand over the past 20 years, anticipating food shortages in their own countries. Why those food shortages? Well, for one their populations are growing fast and out of control; also, industrial development in those countries has been polluting extensive agricultural land areas, with decreasing food outputs each year. Thirdly, these countries anticipate warfare activity among several countries, and in the process increasing land pollution with minerals, as well as contamination of the atmosphere which will contribute to reduced rainfall in those agricultural areas. Results? Food shortages in most countries, especially in the economically poor countries.

Food shortages will be followed by *job shortages* and people desperately trying to *migrate to other countries* in search of jobs and a better life. Migration to other countries is already being restricted seriously, however, as we see it happening in Europe on this second decade of the 21st Century. Already happening? Yes, many of the 27 countries which make up the European Union (EU) today are changing their minds, and refuse to allow emigrants into their countries, as is the case in the United Kingdom, France, Poland, Hungary, and many other countries in the EU. What happens next? *International mafia networks* operate to move emigrants from one country to another, across seas and deserts, across valleys and mountains, charging moneys to desperate families who must endure criminal activity and violation of their basic human rights. We are already seeing these mafia networks operate at multiple points along the Mediterranean Sea, bringing in emigrants from the Middle East and from Africa into countries in the European Union, or at least trying with great losses in human lives.

Similarly, the *United Kingdom* (UK) has decided to leave the European Union (EU) in order to avoid –among many reasons – receiving thousands of more emigrants into its cities, to avoid having to provide social services for these migrating families. The flow of emigrants from *South America to the USA* is now also being obstructed by the new regulations imposed by the new President Donald Trump and his hand-chosen administration, as *Figure 1* shows. Migration into the USA from many Latin American countries has been happening for many decades, coming along a long string of cities all the way from Argentina, to Chile, Venezuela, Guatemala, and Mexico, all the way up to Rio Bravo, the frontier between Mexico in the south and the USA in the north. We know that these emigrant families have contributed greatly to the economy, culture, and political world of the USA, but the times and the people are changing, with large populations in some cities and few jobs available.

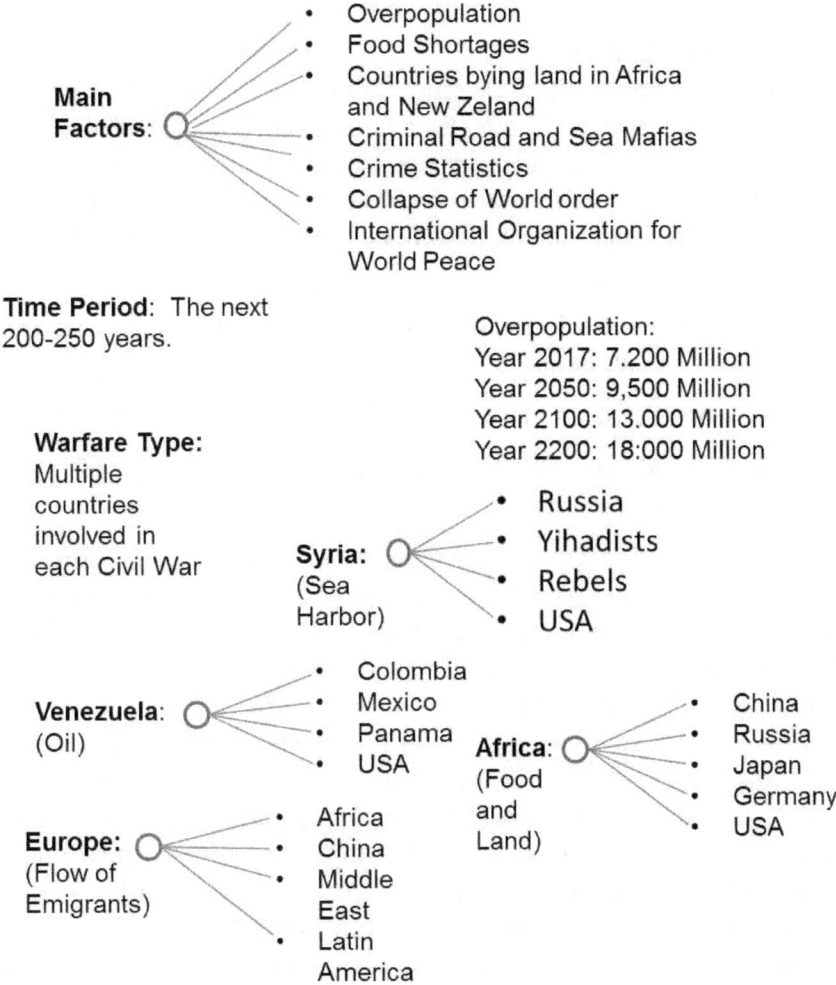

Main Factors: O
- Overpopulation
- Food Shortages
- Countries bying land in Africa and New Zeland
- Criminal Road and Sea Mafias
- Crime Statistics
- Collapse of World order
- International Organization for World Peace

Time Period: The next 200-250 years.

Overpopulation:
Year 2017: 7.200 Million
Year 2050: 9,500 Million
Year 2100: 13.000 Million
Year 2200: 18:000 Million

Warfare Type: Multiple countries involved in each Civil War

Syria: O (Sea Harbor)
- Russia
- Yihadists
- Rebels
- USA

Venezuela: O (Oil)
- Colombia
- Mexico
- Panama
- USA

Africa: O (Food and Land)
- China
- Russia
- Japan
- Germany
- USA

Europe: O (Flow of Emigrants)
- Africa
- China
- Middle East
- Latin America

Figure 1. Scenario 1: Overpopulation, Food Shortages, and Multiple Civil Wars (Part 1 of 2)

Another half hour goes by, with a busy question-and-answer session. Each time the questions get deeper into the reasons for contemplating a scenario such as this Scenario 1.

--OK, as I promised earlier, my partner in this project, Xabier Elurmendi, will now take over the presentation. Please feel free to ask questions along his talk.

Xabier Elurmendi

--Good afternoon, Ladies and Gentlemen. As Kathy already mentioned my name is Xabier Elurmendi, and I will be conducting the second half of this presentation on Scenario 1. My part of the presentation deals with some of the wars and guerrilla activities which we believe could very well happen over the next 200-250 years.

Before Xabier is able to continue he sees another hand rise up in the air among the audience.

--Yes, your question please.

--What would you say are the causes for these wars, and who is participating in these wars?

--Good question. Oil shortages in particular, and alternative energy shortages in general is the answer. Yes, as the population continues to grow unchecked, we will be using more of the oil and gas reserves left within a few countries like Saudi Arabia, Venezuela, Irak, and Russia, and the struggle for control of those oil and gas reserves will increase resulting in wars. Many of us in the IWP team believe the war in *Syria*, for example, will continue among the participating countries and their armed forces, namely the Syrian government forces, Russia's air forces, the Jihadist terrorist armies, the Rebel armies, and occasionally the USA naval and air forces. Many interests and many armed forces come into play. Russia wants to continue having access to one of Syria's naval ports, as to maintain its presence in the Mediterranean Sea, for one. The Jihadists want to expand the powers of its own Caliphate; the Rebel forces oppose the government forces of the Syrian government led by President Bashar al-Assad; and the USA forces are participating in an attempt to control terrorist activity in that area and other parts of the world. Oil? Yes, of course, all these interacting forces want to be close to oil sources in Syria, as well, as we show on *Figure 2*.

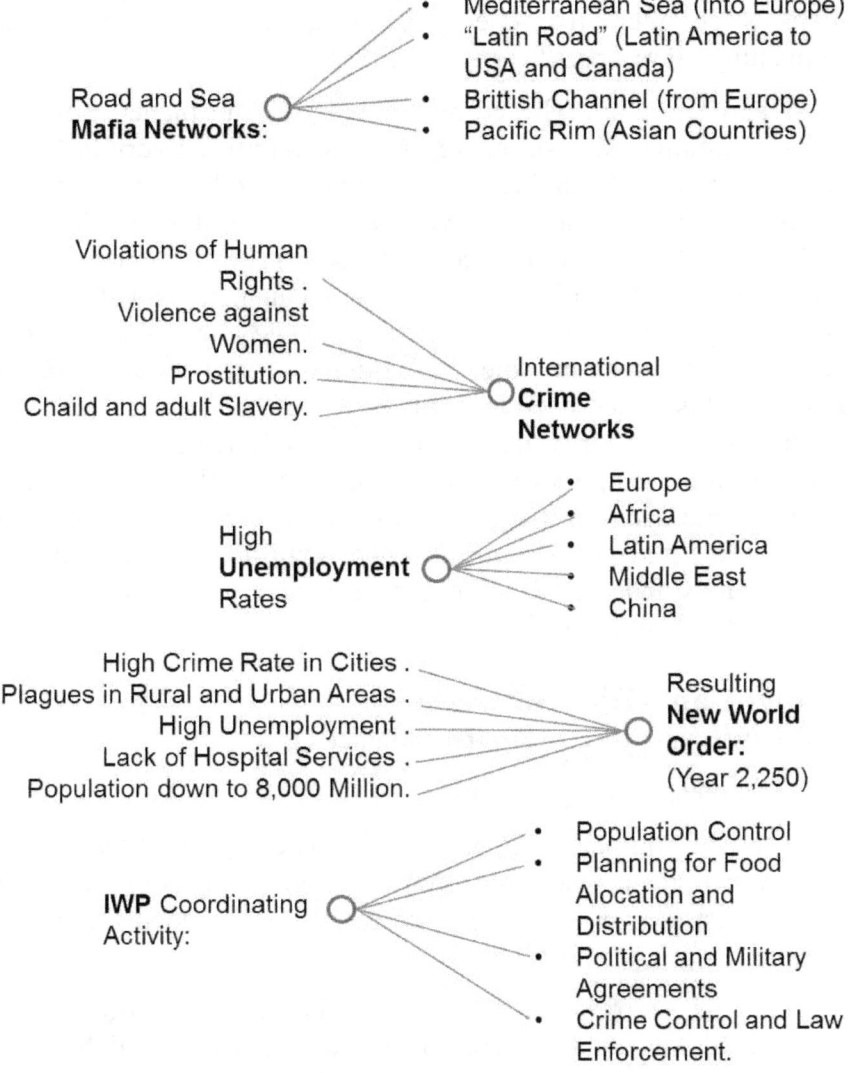

Road and Sea Mafia Networks:
- Mediterranean Sea (into Europe)
- "Latin Road" (Latin America to USA and Canada)
- Brittish Channel (from Europe)
- Pacific Rim (Asian Countries)

Violations of Human Rights .
Violence against Women.
Prostitution.
Chaild and adult Slavery.
→ International **Crime Networks**

High **Unemployment** Rates
- Europe
- Africa
- Latin America
- Middle East
- China

High Crime Rate in Cities .
Plagues in Rural and Urban Areas .
High Unemployment .
Lack of Hospital Services .
Population down to 8,000 Million.
→ Resulting **New World Order:** (Year 2,250)

IWP Coordinating Activity:
- Population Control
- Planning for Food Alocation and Distribution
- Political and Military Agreements
- Crime Control and Law Enforcement.

Figure 2. **Scenario 1:** Overpopulation, Food Shortages, and Multiple Civil Wars (Part 2 of 2)

--But oil reserves are quickly being used everywhere. In fact, oil reserves may only last for another 30-50 years. So, why not place emphasis on alternative energy sources, instead of places with diminishing oil reserves?

--I agree, totally. The current powers in the planet, however, want to continue using oil energy resources until its very last drop, given that much of our industrial complexes today run on oil and gas energy sources, mainly. *Venezuela* – Xabier quickly added – is another country where the conflict may grow to much greater proportions, with Colombia, Mexico, Panama, and the USA participating in order to insure their influence and control over oil resources in such Latin American country. Yes, **President Nicolas Maduro**, his vice-president *Tareck El Aissami*, and his administration may not have much time left to govern, and the hostilities may temporarily subside, but the conflict will reappear time and time again in that part of the globe, we believe. There are several other parts of the world where conflicts over oil and gas resources may and will occur, as I mentioned earlier, such as Russia, Alaska, Africa, and Latin America.

A "Fight-and-Die" Scenario
Another hand is raised in the audience.

--Yes, please, your question.

--Why do you place this scenario in the category of "***fight-and-die***"? Are there no ways of conflicting sides and countries being able to consider matters and come to an understanding, for example?

--In this extreme Scenario 1, the local, national, and international law enforcement forces and networks will be over-extended, not being able to cope with violence, crime, and violation of human rights in every part of the planet. As a result, large segments of the populations in many countries will be involved in crime mafia networks, prostitution for both women and men, plus child and adult slavery. Slavery? Yes, slavery, as many men and women search for food and shelter protection. Hospital networks and private health services also will be over-extended, unable to provide for the needed health services to those 18,000 Million people projected over the next 200-250 years. Yes, periods of disease, poverty, warfare, and plagues will interchange with brief 50-70 year periods of recuperation and "peace", *a cycle to repeat itself over and over, many times*, approaching the collapse of the human species in this planet Earth beyond the year 2200, of course. Sorry, but that is how our IWP team sees this particular Scenario 1.

Role of the IWP Organization

--What can international organizations, such as the IWP, do to try to prevent such an extreme scenario?

--That is also a very good question, one which would be best answered by *Dr. Eugene Finley*, here with us today. Dr. Finley?

--Thank you, Xabier and Kathy! Yes, I am very pleased to be here today, and I will try and answer that question, of course. As you all may know by now, our organization, the *International Organization for World Peace* (IWP), is different from many other international organizations, in many respects. The United Nations (UN) and other international organizations, for example, have a mandate that requires they make their presence *after* a conflict has occurred, be it a war, an environmental disaster, or a plague. By contrast, our organization IWP has a mandate which requires to monitor places in the planet, so that when a list of conditions falls below an established level, such as poverty level, crime level, food shortage, an imminent war conflict, etc., we act and participate with the powers and sides in conflict to try to prevent an outbreak of war or environmental disaster. This being our purpose, we participate with a large number of countries in population control, we give advice and lend human resources to prevent food shortages, we provide support to law enforcement agencies at the national and international levels, plus we hold conferences to encourage governments to anticipate conflicts and to provide resources to avoid such conflicts in the near future.

Ahh, yes, one more thing, please complete the one-page forms available on your tables with recommendations for IWP support to the topics discussed in this Scenario 1.

Again, many thanks to all of you, Ladies and Gentlemen.

Chapter 6:
Scenario 2: Control of Energy Sources, Guerrilla Warfare across the Planet

*"The **human overpopulation issue** is the topic I see as the most vital to solve, if our children and grandchildren are to have a good quality of life."*

-- *Alexandra Paul*
Read more at:
https://www.brainyquote.com/quotes/keywords/overpopulation.html

*"We already have the statistics for the future: the growth percentages of pollution, **overpopulation**, and desertification. The future is already in place."*

-- *Gunter Grass*
Read more at:
https://www.brainyquote.com/quotes/keywords/overpopulation.html

*"Futurists don't consider overpopulation one of the issues of the future. **They consider it the issue of the future**."*

*-- **Dan Brown***
Read more at:
https://www.brainyquote.com/quotes/keywords/overpopulation.html

Xabier Elurmendi

--Good afternoon everyone, Ladies and Gentlemen! I would like to thank everyone for being here today, for joining us in this Seminar series at the ***International Organization for World Peace (IWP)****,* here in Tucson, Arizona, USA. My name is ***Xabier Elurmendi***, a member of IWP team under the direction of ***Dr. Eugene Finley***. This time it is my turn to present the first half of *Scenario 2* with title:

"Control of Energy Sources, Guerrilla Warfare, and Large Migration Flows"

For the second half of the presentation my partner ***Ms. Kathy Thompson*** will take over and she will also help with questions that you may have, OK? As you may recall, when we presented Scenario 1 last week, the main topic was ***Overpopulation*** in our planet and all the associated problems with such situation. It was a very successful meeting we believe, with many good questions, and recommendations. By-the-way, we would like to thank all those folks who later contributed to our *Scenario Series* with recommendations to the IWP, recommendations on how to plan and prepare for those extreme events in the next few decades. We are very grateful, indeed.

Main Factors

A hand is raised in the audience.

--Yes, please, what is your question?

--What is a main theme for this Scenario 2, that is, how is it different from the other scenarios?

--Good question. The key words and main theme for this Scenario 2 would have to be *"Energy Sources in the planet and their control."* Yes, we are going to be talking about the types of energy sources available, the current use by country on a daily basis of those energies, and the warfare activity which we believe is going to happen as the large economies try to secure and own those energy sources. In fact, a listing of the *main factors* in this Scenario 2 is a follows:

- **Limited Oil Resources**
- **Declining global Economies**
- **Oil producing countries agree to increase oil prices**
- **Warm temperatures throughout the Planet.**
- **International war conflict**
- **Arms traffic networks, and**
- **Collapse of the World Order**

In order to go over these main factors and present statistics, we are going to need some 35-40 PowerPoint slides, approximately, so I will present the first half of these slides, and next Kathy will follow with the rest. Please feel free to ask questions at any time, really.

Time Frame

--You may have already mentioned this, but what is the *time period* that we are talking about? Is this scenario to happen sometime in the next 200-300 years, in the next 1,000 years? What is your assessment?

--Yes, I hear you. We are talking about the next 250 years beginning on this second decade of the 21st Century. Yes, things are happening very fast, really. We may have oil resources for the next 30-50 years, and that is it. We better prepare and continue developing alternative energy sources like solar energy and wind energy in order to provide for needs in the centuries ahead. Meantime we are going to have many wars among countries which are oil producers and countries which are oil consumers. In fact, let us now take a look at *Figure 1* and *Figure 2* in the next PowerPoint slides. We observe that Russia is currently the largest oil producer with a total of 10,250,000 barrels/day, followed by Saudi Arabia with 10,050,000 barrels/day, and the United States with 8,744,000 barrels/day. A synopsis of this Scenario 2 is presented in *Figure 3.*

Country	Production (Barrels/Day)
1 Russia	10,250,000
2 Saudi Arabia (OPEC)	10,050,000
3 United States	8,744,000
4 Irak (OPEC)	4,836,000
5 People's Rep of China	3,938,000
6 Iran (OPEC)	3,920,000
7 Canada	3,893,000
8 United Arab E. (OPEC)	3,188,000
9 Kuwait (OPEC)	3,000,000
10 Brazil	2,624,000
11 Venezuela (OPEC)	2,316,000
12 Mexico	2,193,000
13 Norway	1,763,000
14 Kazakhstan	1,746,000
15 Nigeria (OPEC)	1,476,000
16 Angola (OPEC)	1,507,000
17 Algeria (OPEC)	1,171,000
18 United Kingdom	978000
19 Colombia	955000
20 Azerbaijan	876000
21 Indonesia	847000
22 India	736000
23 Malaysia	668000
24 Qatar (OPEC)	639
25 Egypt	582000

Figure 1. Top 25 Oil-producing Countries.[3]

Country	Production (Barrels/ Day)
26 Ecuador (OPEC)	555000
27 Argentina	536000
28 Libya (OPEC)	528000
29 Republic of Congo	317000
30 Vietnam	312000
31 Australia	292000
32 Thailand	265000
33 Sudan	255000
34 Turkmenistan	235000
35 Equatorial Guinea	227000
36 Gabon (OPEC)	210000
37 Denmark	149000
38 Chad	115000
39 Brunei	113000
40 Pakistan	96600
41 Italy	90000
42 Uzbekistan	85000
43 Cameroon	81000
44 Romania	80000
45 South Korea	79000
46 Timor-Leste	76000
47 Trinidad	75000
48 Bolivia	67000
49 Ukraine	66000
50 Bahrain	64000

Figure 2. Top 26-50 Oil-producing Countries.[3]

OPEC Members

Among the members of the *Organization for Petroleum Exporting Countries (OPEC)* we find: Saudi Arabia, Irak, Iran,

United Arab Emirates, Kuwait, Venezuela, Nigeria, Algeria, Angola, Qatar, Ecuador, Libya, and Gabon.

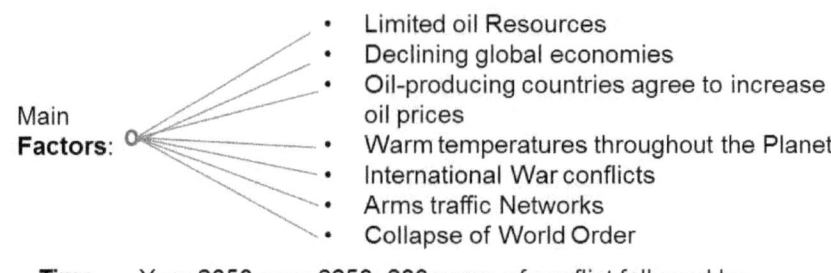

Main **Factors**:
- Limited oil Resources
- Declining global economies
- Oil-producing countries agree to increase oil prices
- Warm temperatures throughout the Planet
- International War conflicts
- Arms traffic Networks
- Collapse of World Order

Time Period: Year 2050-year 2250; 200 years of conflict followed by 50-70 years of recuperation and "peace"

Warfare Type: Coalitions of 3-5 countries make war on oil producing countries.

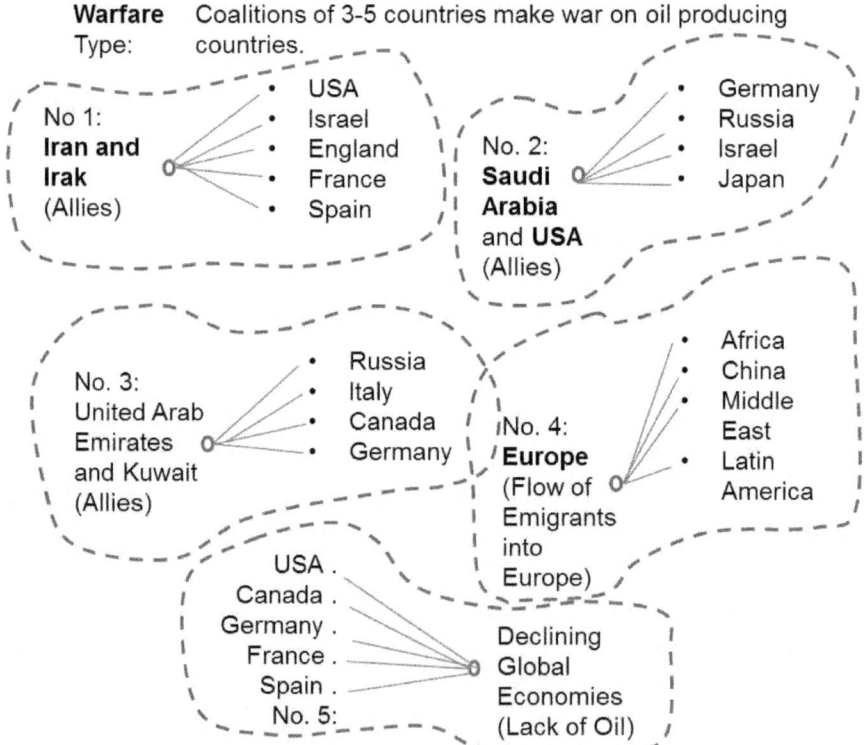

***Figure 3*. Scenario 2**: Control of Energy Sources, Guerrilla Warfare, and Large Migration Flows. (Part 1 of 2)

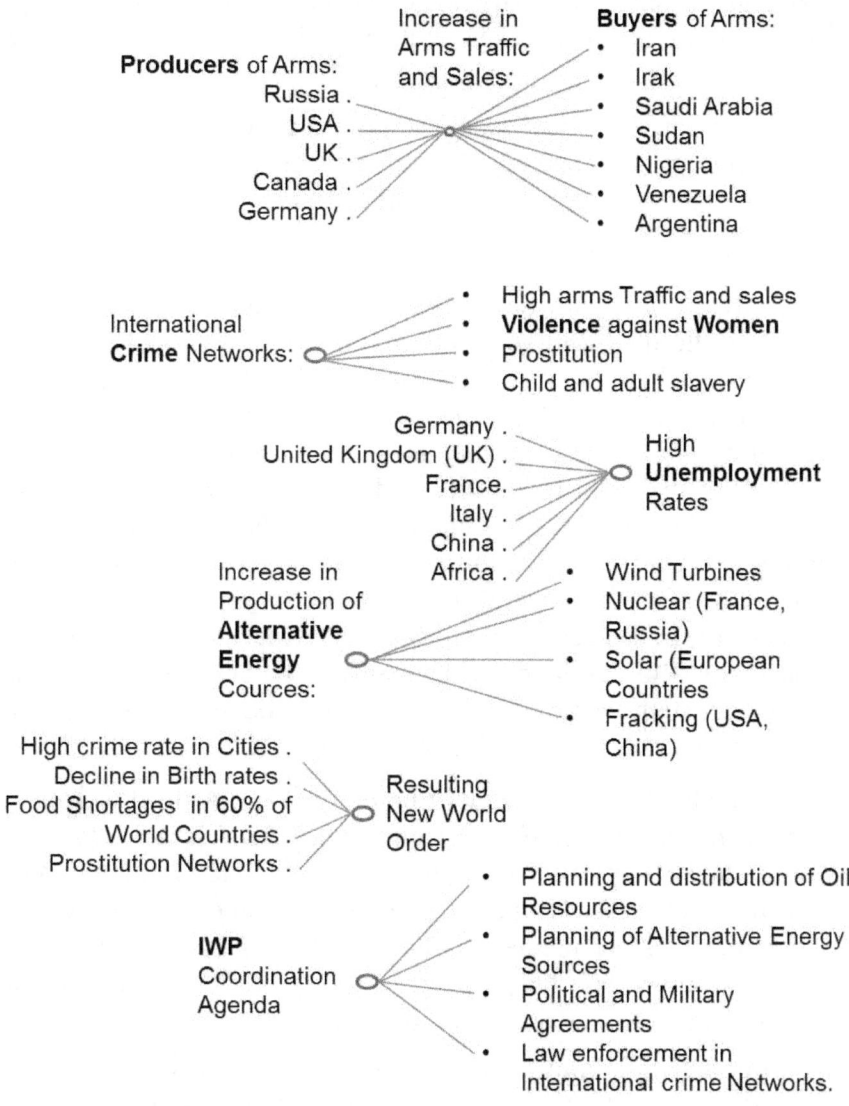

Figure 3. **Scenario 2**: Control of Energy Sources, Guerrilla Warfare, and Large Migration Flows. (Part 2 of 2)

--Yes, your question, please.

--But if we have so many countries producing oil, up to 70 Million barrels a day, why does your team contemplate a long list of wars among oil-producing countries and oil-consumer countries?

--Because oil resources are just that, limited in their yield and can only last so many years, 30-50 years, we estimate. These wars may emerge as guerrilla warfare with only 2-3 countries involved, but eventually more countries will become involved, we believe.

--For example?

--Well, our team believes that it is realistic to envision 4 such wars in the next 250 years beginning with *Iran and Irak as allies*, defending their oil resources, with USA, Israel, England, France, and Spain on the other side, at first trying to work out a buying program, and later, in the absence of any deal, the two sides engaging in guerrilla activity and eventual all-out wars. Similarly, our team believes that *Saudi Arabia and the USA* will end up being allies as they defend their oil resources and interests against Germany, Russia, Israel, and Japan who will be competing to gain access to the oil resources in Saudi Arabia. We are not saying that Germany, Russia, Israel, and Japan will be entering an agreement to carry out such war, no, and instead we believe they will act on their own, each one, individually, and not as a pact.

We also believe there will be other wars. A third war being that between *United Arab Emirates and Kuwait as allies* against individual efforts by Russia, Italy, Canada, and Germany. Why not? Currently Russia is the largest oil producer in the world, but potential mismanagement of its oil resources in the near future would venture this large country into seeking partial ownership of oil resources and reserves in other countries.

Of course, one main activity such as oil-based wars cannot remain as the single focus of attention. We also believe that these oil-based wars will also trigger the purchase of arms and the flow of emigrants at very high international levels. We envision for, example, a *very high traffic of arms* between weapon-producing countries, such as Russia, USA, UK, Canada, and Germany, and weapon-buying countries like Iran, Iraq, Saudi Arabia, Sudan, Nigeria, Venezuela, and Argentina. Oil-producing countries will have the option of either investing in technologies and making their own weapons, or buying those weapons; they will opt for buying those weapons and then concentrate on their oil-producing empires.

Alternative Energy Sources

A hand rises in the audience.

--I hear you, but what happens if the production of *alternative energy sources like solar energy, wind energy, gas reserves, and hydroelectric power* begins to take a more important role in the planet and *the demand for oil decreases*?

--Ahh, that is a very interesting question. In fact, we may want to ask you to join our team at IWP so you can help us design the other scenarios!

Applause in the audience.

--But at this point, it is time for my partner in the IWP team, Ms. *Kathy Thompson*, to join us and help with the second half of this presentation. Kathy, could you join us?

Kathy Thompson

--Thank you, Xabier. Hello everyone, my name is Kathy Thompson, also a member of the IWP team as Xabier has mentioned. Yes, as our guest has indicated, we may also encounter a decrease in the demand for oil resources due to the new alternative energy sources, but that is a topic which we are going to address in the following few weeks within other scenarios. For this Scenario 2 another by-product will be the *high flow of emigrants into Europe* and coming in from many countries and continents, including Africa, China, Middle East, and Latin America. Already now, with the Syrian war going on, on this 2nd decade of the 21st Century, there is a large number of emigrants from a long list of countries trying to enter Europe to escape devastation and death, in search of a better life, as shown on *Figure 4*.

Flow of Emigrants

Countries with a large percentage of immigrants are Russia, Germany, and United Kingdom with 7.7%, 11.9%, and 12.4% respectively. Of course, we also have Switzerland with a high percentage of 28.9%, as shown on *Figure 4.*

Country	Number of Immigrants	Immigrants as Percentage of National Population (%)
1 Russia	11,048,064	7.7
2 Germany	9,845,244	11.9
3 United Kingdom	7,824,131	12.4
4 France	7,439,086	11.6
5 Spain	6,466,605	13.8
6 Italy	5,721,457	9.4
7 Ukraine	5,151,378	11.4
8 Switzerland	2,335,059	28.9
9 Netherlands	1,964,922	11.7
10 Sweden	1,130.025	15.9
11 Austria	1,333,807	15.7
12 Belgium	1,159,801	10.4
13 Belarus	1,085,396	11.6
14 Greece	988,285	8.9
15 Portugal	893,847	8.4
16 Croatia	756,98	17.6
17 Ireland	735,535	15.9
18 Norway	694,508	13.8
19 Poland	663,755	0.9
20 Denmark	556,825	9.9
21 Serbia	532,457	5.6
22 Hungary	449,632	4.7
23 Czech Republic	439,116	4.0
24 Moldova	391,508	11.2
25 Azerbaijan	323,843	3.4
26 Armenia	317,001	10.6
27 Finland	293,167	5.4
28 Latvia	282,887	13.8
29 Slovenia	233,293	11.3
30 Gibraltar	9,662	33.0

Figure 4. List of Top 30 countries with Immigrant Population. [5]

Saturation of the economic markets will also become a main factor in this Scenario 2, we believe. In 70 years of relative peace since the end of hostilities in World War II, we have managed to manufacture and sell to every country in the planet astronomical numbers of refrigerators, washing machines, television sets, furniture, air conditioning systems, automobiles, homes, and hundreds of other products, indeed. Europe is practically saturated with all these products, and so are many other countries in Asia; by contrast, countries in the African continent remain in deep poverty. If we were smart enough in the Western World, we would help Africa achieve higher economic means so that it could afford to buy goods from Europe, extending the period of productivity and relative wealth in this continent. While Africa remains poor, the flow of emigrants to the European continent will continue, unable to buy goods from the same European continent. As a result, the *economies of many countries will enter a declining phase*, as will be the case with USA, Canada, Germany, France, Spain, China, and many other countries.

Mafia Networks

--Ms. Thompson, could you please comment on the *quality of life* over the next 250 years?

--Certainly, and wish I could be positive about the prospects of a good life over that time period, but *crime will be on the rise*, I'm afraid. With all the guerrilla and war activity which we have mentioned, the large flows of emigrants from some countries to other countries, and the declining economies. There will always be international organizations with large economic powers and fiscal paradises for a few individuals, but the majority of the population will begin to experience crime at several dimensions and levels, including large mafia networks in cities, violence against women, highly visible women and men prostitution networks, as well as child and adult slavery networks.

Back in the audience another hand raises in the air.

--Ms. Thompson, you have painted a very dark and grim scenario awaiting all of us, terrifying I might add. Will it mean the end of the world as we know it today?

--Not so, but close. "The end of the world" are big words, very big, but we human beings are very durable, very resilient, and we would continue to live for a few more hundreds of years. The scenario we have painted is one where humanity falls into a 250-year period of piracy, theft, high crime, death by the millions, and low economic productivity. Next will follow a time period of 50-70 years of relative calm and recuperation. That is the time cycle which we paint in this Scenario 2: a period of 250 years of wars and death, followed by a time period of 50-70 years of recuperation. At the end of such time period of recuperation, *the cycle begins all over again*. How many times can this cycle repeat itself before total auto-destruction of the human species, or before we enter a different scenario? We just don't know.

--So what can we do as a society today to try and prevent such *Scenario 2* as your team has presented?

--Again, for such a question, I would ask our team leader, *Dr. Eugene Finley*, to participate and to share with us some of the planning already taking place within the IWP organization. Dr. Finley?

--Thank you, Kathy. We are already working with some 25-30 countries to promote the development of renewable energy resources, including *solar energy, wind energy, and hydroelectric power*, to mention the 3 sources most important to us in the IWP organization. We have already discarded *nuclear power* as an option because of the highly costly radioactive consequences of such a power. We are also *promoting the availability of a "global energy market"* with reasonable access and reasonable energy prices, which means that energy buyers would be invited to invest capital and manpower in the energy-producing countries as well.

--And what about the large migration flows that we are experiencing in this 21st Century across the five continents?

--My own personal insight into this matter is that the *emigration flows will continue*, and will become part of our new planet Earth. People will move, displace themselves, and emigrate to better places in search of a better life. Now days you can visit almost any city in the world and you will see people from dozens and hundreds of other countries, working in kitchens, selling a variety of products on street carts, laying bricks, cleaning windows, etc. Our world has already changed, whether we like it or not, and our cities are already multi-racial and multi-cultural, offering goods and services from all over the planet. By-the-way, please remember to fill those 1-page *survey sheets* on your tables with recommendations to the IWP organization, one that we all belong to.

Again, many thanks to each and everyone of you for joining us in the presentation of *Scenario 2*.

Chapter 7:
Scenario 3: Climate Change and Environmental Holocaust

*"The damage that **climate change** is causing and that will get worse if we fail to act goes beyond the hundreds of thousands of lives, homes and businesses lost, **ecosystems destroyed,** species driven to extinction, infrastructure smashed and people inconvenienced."*
-- David Suzuki
Read more at:
https://www.brainyquote.com/quotes/keywords/climate_change.html

*"Some argue that now isn't the time to push the green agenda, that all efforts should be on preventing a serious recession. That is a false choice. It fails to recognize that **climate change and our carbon reliance is part of problem** - high fuel prices and food shortages*

due to poor crop yields compound today's financial difficulties."
--Lucy Powell
Read more at:
https://www.brainyquote.com/quotes/keyword s/climate_change.html

--Welcome to our 3rd Seminar at the **International Organization for World Peace (IWP)** here in our favorite City, Tucson, Arizona, USA. My name is **Xabier Elurmendi**, a member of the IWP, and this time the title of our presentation is:

"Climatic Change and Environmental Holocaust"

As the title says, our main topic for this afternoon is *"climatic change"*, to be followed by a list of extreme global conditions associated with such change. Also, before I forget, I want to thank all of you for attending the 2nd Seminar last week, and for handing in your individual lists with recommendations to the IWP organization; wonderful ideas on how to prepare for the difficult decades ahead of us, as individuals but also as a society, as a community of nations in our planet Earth.

--Any questions from anyone in the audience before I begin this presentation?

Main Factors

A hand is seen up in the air on the first row of tables in the conference room.

--Yes, please go ahead with your question, Madam.

--Thank you. If you will, please let us know about that list of extreme global conditions that distinguishes this Seminar from the other seminar and, in your opinion, what is the probability that this scenario will take place in the next few years or next one hundred years.

--Certainly, will be glad to do so, of course. A list of the **main factors**, the main players, extreme conditions in this **Scenario 3** are as follows:

- Climatic change (C.C.) and **gradual warming** of the Earth by an extra 2° Celsius (3.6° Fahrenheit).

- **Death toll** due to climatic change of 300,000 lives/year.

- **Hunger** due to Climatic change, up to 45 Million people.

- **Global costs** due to climatic change of up to $125,000/year,

- And up to **1/4 of the Earth's plants (flora) under threat of death** due to climatic change, especially in Brazil, Peru, and Colombia, central Africa, and central Asia.

With regards to your second question, the probability of occurrence of this event, climatic change, I would have to say that it is **100% probable**, that is, it is already happening. Of course, I would have to add that **these scenarios are not "mutually exclusive"** so that if one scenario occurs then one or more of the other scenarios can occur as well. Of the 8 scenarios that we anticipate, at any time we may end up having 2-3 scenarios happening simultaneously, which complicates matter altogether.

Climate change

Climate change being the key factor in this **Scenario 3**, I would then suggest we take a look at the following PowerPoints with these details:

(PowerPoint slide No. 3):
More than 300 million people are already seriously affected by the **gradual warming of the earth** and that number is set to double by 2030, the report from the *Global Humanitarian Forum* warns. "*For the first time we are trying to get the world's attention to the fact that climate change is not something waiting to happen. It is impacting seriously the lives of many people around the world,*" the

forum's president, former *U.N.* Secretary-General *Kofi Annan*, told CNN.

Speaking to CNN's *Becky Anderson* in London on Friday, *Annan* said the migration of people from newly uninhabitable areas presents a security issue that needs to be addressed by the United Nations Security Council. *"This is one of the reasons why I've described* **climate change** *as all encompassing,"* he told CNN. *"This threat to our health, this threat to food production, this threat to security. It raises political tensions, it will have people on the move -- and they are on the move -- and many more which will bring tensions."*

The report, titled *"**Human Impact Report: Climate Change -- The Anatomy of a Silent Crisis**"* comes just six months before the *United Nations Climate Conference* in Copenhagen to forge a post-Kyoto climate agreement for 2012 and beyond. Annan called on Member States to reach a "global, effective, fair and binding" outcome on climate change, as the report warned that the talks could "well be the last chance for avoiding global catastrophe." He told CNN: *"The U.S. administration has joined the mainstream about fighting climate change and that is a big step, and I hope that will also put a new momentum into the negotiations."* The report's startling numbers are based on calculations by the ***Intergovernmental Panel on Climate Change*** that the Earth's atmosphere warmed by 0.74 degrees Celsius (1.33 degrees Fahrenheit) from 1906 to 2005, with much of that increase coming in recent decades. The panel predicts that ***by 2100*** temperatures will have increased a minimum of ***two degrees Celsius*** (3.6 degrees Fahrenheit) over pre-industrial levels regardless of what's agreed in Copenhagen. "No matter what," the report concludes, *"the suffering documented in this report is only the beginning."* A rise of two degrees, it says, ***"would be catastrophic."***

(PowerPoint No. 4):
Of the ***300,000 lives being lost each year*** due to climate change, the report finds nine out of 10 are related to

"gradual environmental degradation," and that deaths caused by climate-related malnutrition, diarrhea and malaria outnumber direct fatalities from weather-related disasters. The vast majority of deaths – 99% -- are in developing countries which are estimated to have contributed less than one percent of the world's total carbon emissions. The report warns climate change threatens all eight of the Millennium Development Goals-- a set of goals agreed on by leading nations in 2000 that aim to reduce extreme poverty by 2015. The goals include eradicating hunger, reducing child mortality, and halting the spread of diseases including HIV/AIDS and malaria.

(PowerPoint No. 5):
Around 45 million of the 900 million people estimated to be chronically hungry are suffering due to *climate change*, the report says. Within 20 years that number is expected to double. At the same time food production is expected to fall, driving food prices up 20 percent. The countries considered to be most vulnerable are those in the semi-arid dry land belt that runs from the Sahara/Sahel to the Middle East and Central Asia, Sub-Saharan Africa, South and Southeast Asia, Latin America and parts of the U.S., small island states and the Arctic region. Australia is singled out as the developed country most vulnerable to the direct impacts of climate change. Over the past 15 years, the combination of rising temperature and lower rainfall has produced the worst drought in the country's recorded history.

While developed countries -- including Australia -- have committed funds to counter the impact of climate change, the Global Humanitarian Forum says developing nations need a dramatic injection of funds -- up to 100 times more than is currently available to help them adapt to the changes. The total economic cost of climate change each year is thought to *be $125 billion,* although the Forum warns that figure may be too conservative and doesn't take into account the impacts on "health, water supply and other shocks."[13]

--Yes, I've heard of similar reports on my trips to conferences in the Middle East. It is happening now. We're inside of it. But what is the perception of the IWP regarding the duration of this "global warming"? Will stay here forever, will it go away sometime in the future?

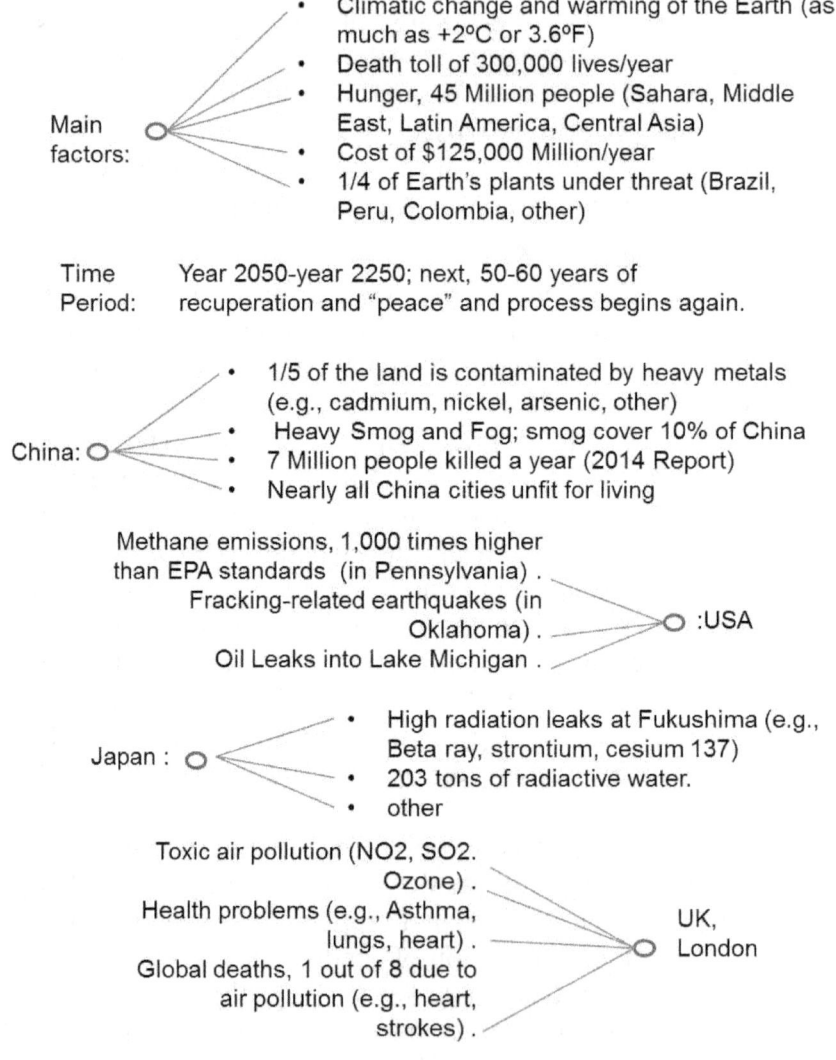

Main factors:
- Climatic change and warming of the Earth (as much as +2°C or 3.6°F)
- Death toll of 300,000 lives/year
- Hunger, 45 Million people (Sahara, Middle East, Latin America, Central Asia)
- Cost of $125,000 Million/year
- 1/4 of Earth's plants under threat (Brazil, Peru, Colombia, other)

Time Period: Year 2050-year 2250; next, 50-60 years of recuperation and "peace" and process begins again.

China:
- 1/5 of the land is contaminated by heavy metals (e.g., cadmium, nickel, arsenic, other)
- Heavy Smog and Fog; smog cover 10% of China
- 7 Million people killed a year (2014 Report)
- Nearly all China cities unfit for living

Methane emissions, 1,000 times higher than EPA standards (in Pennsylvania).
Fracking-related earthquakes (in Oklahoma).
Oil Leaks into Lake Michigan. :USA

Japan :
- High radiation leaks at Fukushima (e.g., Beta ray, strontium, cesium 137)
- 203 tons of radiactive water.
- other

Toxic air pollution (NO2, SO2. Ozone).
Health problems (e.g., Asthma, lungs, heart).
Global deaths, 1 out of 8 due to air pollution (e.g., heart, strokes).
UK, London

Figure 5. **Scenario 3**: Climatic Change and Environmental Holocaust. (Part 1 of 2)

--I'm afraid this Scenario 3 will probably stay with us for many years, possibly hundreds of years, given the continued mining and burning of fuels in our planet. Our team feels that it is on right now, that many negative environmental impacts will take place, and that populations in our cities will also be castigated all over the planet. Again, the cycle is probably like this: Our nations around the globe will continue with mining, burning oil and gas fuels for the next 200-250 years, until we reach a point where we cannot take it anymore and then things will slow down, say for the next 50-70 years to recuperate our population. After those 50-70 years the world nations will go back to activities which will continue to pollute the land and the air, and the cycle will repeat itself again, and again. But let us take a look at some of the statistics already being reported by many international organizations.

(PowerPoint No. 6):
Amazon Forests. Amazonian forests have lost about 12% of their original extent and are projected to lose another 9 to 28% by 2050. The consequences of ongoing forest loss in Amazonia (here all rainforests of the Amazon basin and Guiana Shield) are relatively well understood at the ecosystem level, where they include soil erosion, diminished ecosystem services, altered climatic patterns, and habitat degradation. By contrast, little is known about how historical forest loss has affected the population sizes of plant and animal species in the basin and how ongoing deforestation will affect these populations in the future.

As a result, the conservation status of the 15,000 species of plants that compose the Amazonian tree flora—one of the most diverse plant communities on Earth—remains unknown. To date, only a tiny proportion of Amazonian tree species have been formally assessed for the ***International Union for Conservation of Nature*** (IUCN) Red List. Two previous studies have attempted to estimate the extinction threat to Amazonian plants using theory, data, and vegetation maps to model reductions in range size, but they disagreed on whether the proportion of threatened plant species in the Amazon is low (5 to 9%) or moderate (20 to 33%).

Here, we build on that work by using a spatially explicit model of tree species abundance based on 1485 forest inventories to quantify how historical deforestation across Amazonia has reduced the population sizes of 4,953 relatively common tree species. We use a separate model to estimate population declines for an additional 10,247 rarer tree species. For both models, we also estimate the population losses expected for 2050 under two deforestation *scenarios* and ask to what extent projected losses can be prevented by Amazonia's existing protected area network. In contrast to previous studies, which presented results in the currency of statistical probability of extinction, we interpret our results using the criteria of the IUCN Red List of Threatened Species, the most commonly used yardstick for species conservation status.[1][2]

Mining and Industrial activity in China

Let us also take a look mining and industrial activity in *mainland China*, as one of the many extreme activities listed earlier.

(PowerPoint No. 7):
Fifth of the farming land in China is contaminated, according to a joint report issued by the country's *Ministry of Environmental Protection and the Ministry of Land and Resources.*

The report, based on surveys conducted between April 2005 and December 2013, found 16.1% of the land including 19.4% of the farmland in the Chinese mainland *contaminated with heavy metals.* Nearly 83 percent of the polluted land is contaminated with *cadmium, nickel and arsenic*. Note: Based on the data available, the *International Agency for Research on Cancer (IARC)* classifies cadmium, nickel and arsenic and their compounds as "carcinogenic to humans." "The general condition of the land is 'not optimistic' as the quality of farming land is worrying and deserted industrial and mining land suffers serious pollution."

The levels of cadmium contamination rose by 50% in southwestern and coastal regions and 10% to 40% in other parts of China between 1986 and 1990, said the report.

Dangerous levels of cadmium were detected in rice produced in central China's Hunan Province, the country's top rice-growing region, which caused a public outcry last year. "Heavy metal pollution alone has resulted in *the loss of 10 million tons of grain and the contamination of another 12 million tons annually, incurring 20 billion yuan (3.17 billion U.S. dollars) in direct economic losses each year,*" said the report citing official estimates. "The main pollution source is human industrial and agricultural activities," the report said. Collection, storage, transfer and disposal of dangerous waste are currently unregulated. "Compared with air and water pollution, *soil pollution* is more difficult to control and remedy, taking a much longer time and needing more resources," according to the Chinese Academy of Sciences.

The report cites the *Dabaoshan coal mine in Shaoguan City* in Guangdong, as an example. Since the 1970s, untreated coal residue and wastewater from the mine have contaminated the soil in the surrounding region up to 44 times above the national limit. "Known to produce some 6,000 tons of copper and 850,000 tons of iron ore annually, the mine has produced a growing amount of sludge and wastewater that has contaminated some 585 hectares along the lower sections of the Hengshui River running atop the mountain." *At least 250 cancer-related deaths associated with soil pollution were recorded in Shangbai village, located downstream of the Dabaoshan mine, between 1987 and 2009.* In 1990, a farmland area of 0.06 hectares near Dabaoshan Mine could yield about 350 kilograms of rice, but by 2010 it could only grow less than 100kg due to heavy soil pollution, said Li Deng'e, a despondent 74-year-old villager, who sprays ever-increasing amounts of pesticides on her land to slow down the daily erosion, reported *China Daily*.[3]

Air pollution

What about China's mechanisms of *air pollution* and its consequences? For that purpose we move to the next PowerPoint slide.

(PowerPoint No. 8):
Satellite remote monitoring showed 980,000km², or 10% of the country, blanketed by smog Sunday including Beijing_and the provinces of Liaoning, Hebei, and Shanxi. Beijing entered its six consecutive day of hazardous smog (Air Quality Index (AQI) above 300), described by as "Apocalyptic smog" by the residents, on Tuesday. However, the authorities have failed to issue a "Red Alert" for the capital, breaking their own rules. "A red alert indicates the most serious air pollution (AQI above 300) for three consecutive days. An orange alert indicates heavy to serious air pollution (AQI between 200 and 300) alternately for three consecutive days. A yellow alert indicates severe pollution for one day or heavy pollution for three consecutive days." Beijing was placed on "Yellow Alert" on Thursday, but the alert was upgraded to "Orange" for the first time on Friday, where it has stayed since. The poisonous smog is forecast to linger until at least Thursday, according to the China Meteorological Administration. PM2.5 AQI for Beijing was 413, indicating a PM2.5 concentration of between 350.5 and 500.4 micrograms per cubic meter, as of posting. Pollution levels below 15.4 micrograms per cubic meter (AQI of up to 50) are considered as "safe."

Chinese Cities Unfit for Living. Nearly all Chinese cities monitored for pollution in 2013 failed to meet environmental standards. *About 96 Chinese cities monitored for pollution in 2013 failed to meet environmental standards, said Wu Xiaoqing, China's vice-minister of environmental protection.* Of the 74 cities monitored by Beijing, 71 had various degrees of problems, said Wu at a news conference on Saturday, according to the report. The only three cities that met the standard were the flood-prone city of Haikou in the island province of Hainan, Llasa, the capital of quake-prone Tibet, and the coastal resort city of Zhoushan. *China's pollution problems* can only be solved through fundamental changes to the way the country develops its economy. "When we were chasing GDP growth, we were also paying the price of pollution, and this price is heavy, is massive." *"China will cut outdated steel production capacity by a total of 27 million tons this year*, slash cement production by 42 million tons and also shut down 50,000 small coal-fired furnaces across the country, according to the

government work report delivered by **_Premier Li_** on Wednesday."
[4]

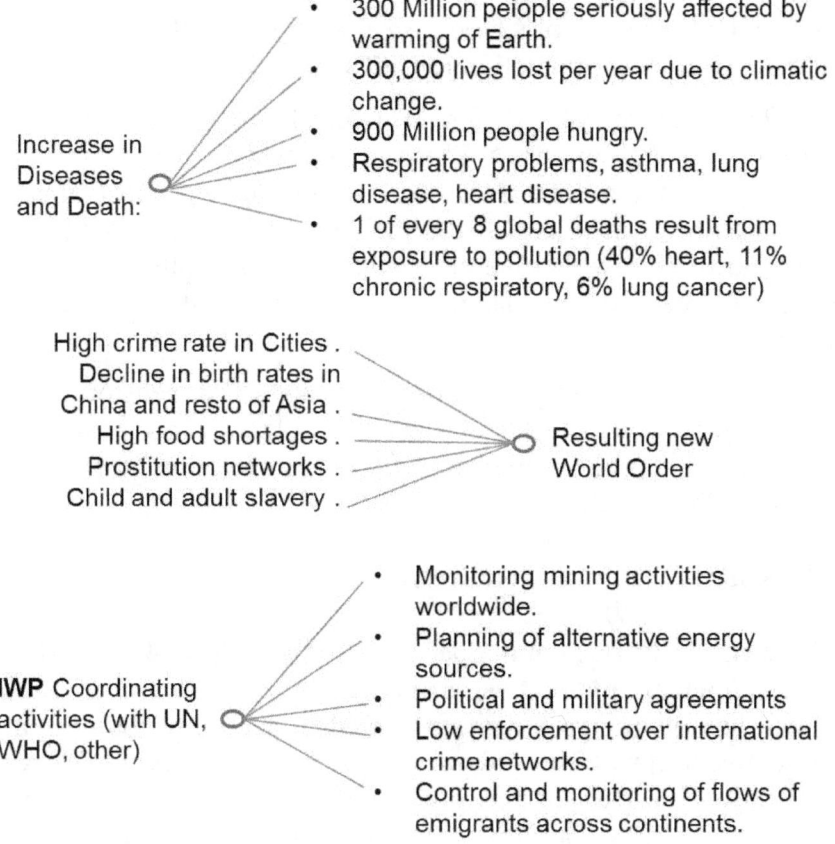

Increase in
Diseases
and Death:

- 300 Million peiople seriously affected by warming of Earth.
- 300,000 lives lost per year due to climatic change.
- 900 Million people hungry.
- Respiratory problems, asthma, lung disease, heart disease.
- 1 of every 8 global deaths result from exposure to pollution (40% heart, 11% chronic respiratory, 6% lung cancer)

High crime rate in Cities .
Decline in birth rates in
China and resto of Asia .
High food shortages .
Prostitution networks .
Child and adult slavery .

Resulting new
World Order

IWP Coordinating
activities (with UN,
WHO, other)

- Monitoring mining activities worldwide.
- Planning of alternative energy sources.
- Political and military agreements
- Low enforcement over international crime networks.
- Control and monitoring of flows of emigrants across continents.

Figure 6. Scenario 3: Climatic Change and Environmental
Holocaust. (Part 2 of 2)

A hand rises in the middle of the conference room.

--If I may, I would like to say that is part of the answer I have been searching for all these weeks regarding the reason for China's *"dumping"* of steel prices in Europe, especially in the Basque Country. Did the price of Chinese steel drop before the decision was made to lower steel production in China?

--Thank you for your comment. Don't know exactly, but that is very possibly the answer to China's "dumping" of steel prices in Europe, yes. While there as an over-supply of steel in China the option was to drop prices in the global market, very possibly. Of course, now we now that the land, water, and air pollution levels in China have reached very dangerous levels for most cities and their population.

Next, we are going to move to another dimension of the "environmental holocaust" that is taking place in our planet, this other dimension being *nuclear radiation*, so we now move to the next PowerPoint slide.

Fukushima Nuclear Incident
(PowerPoint No. 9):
"The criminally negligent operator of the crippled Fukushima Daiichi nuclear plant has admitted that yet another major setback has plagued a key system used to decontaminate highly-radioactive water." The *Tokyo Electric Power Company* (TEPCO) says about 1,100 liters of radioactive water overflowed into a barrier inside the ALPS building. The water was used to wash contaminated equipment and overflowed from a storage facility on Wednesday.

Comedy of Deadly Errors. The workers discovered the leak while cleaning a tank used for filtering radioactive substances from water. The tank is on one of the 3 separate stages of the Advanced Liquid Processing System, or ALPS, local media reported TEPCO as saying.

Figure 3. Photograph shows highly contaminated water leaked from a storage tank at the **Fukushima nuclear power plant in Japan** crippled by a Tsunami.[1]

The operator says the water contains more than 3.8 million becquerels of **beta ray** emitting materials including **strontium** and 6,700 becquerels of **cesium 137**. On Tuesday, TEPCO reported that 203 tons of highly contaminated radioactive water had been pumped into a basement area at the Fukushima in the time period April 10-13. The building is neither a storage area nor a processing facility. The pumps had diverted the contaminated water to the wrong building, the company said. TEPCO says it found four pumps operating at the location that were not even meant to be used. *The water contained radioactive cesium decaying at a rate of 37 million becquerels per liter.*

Some observations from the same report.

1. Radiation dose of about 2,000 millisieverts (200,000 millirems) cause serious illness.

2. Average background radiation in the US is about **3 mS/yr.**

3. The average annual radiation dose per person in the U.S. is currently 620 millirem (**6.2 mSv**), according to EPA. "Half of our average dose comes from natural background sources: cosmic radiation from space, naturally occurring radioactive minerals in the ground and in your body, and from the radioactive gases radon and thoron, which are created when other naturally occurring elements undergo radioactive decay. Another 48% of our dose comes from medical diagnostics and treatments."

Death Statistics

--Those are very good statistics, they point to the reasons our planet may be in the process of becoming an "environmental holocaust", to use your own words, Xabier. But those are statistics about chemical and radioactive agents. How does the proliferation of these agents translate into deaths of people in our planet?

--A very good point to make, and that is why for the 2nd half of our presentation we are going to ask *Ms. Kathy Thompson* to take over. Again, my thanks to all of you for your great questions and comments. Kathy, could you please take over the presentation?

Kathy Thompson

--Thank you Xabier, and thank you to all of you attending this *Seminar 3* on *Climatic Change and Environmental Holocaust*. We are now going to look at air pollution and how it represents a major threat in many cities around the globe. Towards that end, we are going to look at the contents of the next 2 PowerPoint slides with information of air pollution in London, United Kingdom, and Pennsylvania, USA.

(PowerPoint No. 10):
"No other species makes their habitat unlivable! Extremely high levels of air pollution are spreading across parts of England. The pollution is a cocktail of emissions from the UK and Europe, rich in toxic pollutants, including high concentration of the atmospheric particulate matter (PM10, PM2.5), nitrogen dioxide (NO2), sulfur dioxide (SO2) and ozone, mixed with dust from the Sahara—the proverbial icing. The elderly, those with respiratory problems,

asthma, lung or heart disease, have been warned against venturing outside, on the streets.

The latest episode follows the legal proceedings launched against the UK by the **European Commission** for failing to reduce levels of nitrogen dioxide (NO2) in the air after 15 consecutive years of warning, said the report. The head of Asthma UK organization has warned that about 70% of asthma sufferers who find air pollution makes their condition worse "will be at an increased risk of an attack." Even healthy people could experience symptoms including sore throat, eye irritation, nasal discomfort, burning lungs and dry cough.

Figure 4. The City of London, United Kingdom (UK), cloaked in smog.[6]

WHO: Air pollution the world's worst environmental hazard. "1 in 8 of total global deaths occurs as a result of air pollution exposure. This finding more than doubles previous estimates and confirms that **air pollution is now the world's largest single environmental health risk,**" according to the World Health Organization (WHO).

Outdoor air pollution-caused deaths, breakdown by disease:
- 40% – ischaemic heart disease;
- 40% – stroke;
- 11% – chronic obstructive pulmonary disease (COPD);
- 6% – lung cancer; and
- 3% – acute lower respiratory infections in children.

Indoor air pollution-caused deaths, breakdown by disease:
- 34% – stroke;
- 26% – ischaemic heart disease;
- 22% – COPD;
- 12% – acute lower respiratory infections in children; and
- 6% – lung cancer.

"Methane emissions 1,000 higher than Environmental Protection Agency (EPA) estimates. Using an airborne laboratory for atmospheric research, researchers identified and quantified large sources of **methane emissions** over southwestern Pennsylvania, USA, in June 2012. They discovered that emissions rates were up to 1,000 times higher than those estimated by the EPA during the same time period. "We identified a significant regional flux of methane over a large area of **shale gas wells** in southwestern Pennsylvania in the Marcellus formation and further identified several pads with high methane emissions," said the study. "These shale gas pads were identified as in the drilling process, a preproduction stage not previously associated with high methane emissions."

The original sampling area (OSA) encompasses all of Green County, PA, most of Washington County, PA, and parts of Fayette County, PA, Marshall County, WV, and Ohio County, WV, for a total area of 2,844 km², the authors reported. The authors identified **57,673 wells** across the counties of interest. "It is particularly noteworthy that large emissions were measured for wells in the drilling phase, in some cases 100 to 1,000 times greater than the inventory estimates," said one of the report authors. "This indicates that there are processes occurring—e.g. emissions from coal seams during the drilling process—that are not captured in the inventory development process. This is another example pointing to the idea that a large fraction of the total emissions is coming from a small fraction of shale gas production components that are in an anomalous condition." *The comparative impact of methane on climate change is more than 20 times greater than carbon dioxide over a 100-year period, according to EPA.*[7]

A New World Order

--Ms. Thompson, do you really believe that this climatic change will bring *drastic changes to our way of life* in our cities, or will it be something that we will be able to cope with our law enforcement programs?

--Yes, I do. We are already part of this cycle of climatic change today, as we said earlier, and already we are seeing very negative impacts in our cities, such as:

- High crime rates in cities.
- Decline in birth rates in Europe.
- High food shortages over the entire globe.
- High death rates due to a long list of diseases.
- Prostitution networks for women and men controlled by mafia networks, nationally and internationally, and
- Child and adult slavery among emigrant flows of people, from one country to another.

By-the-way, *Dr. Eugene Finley* has not been able to join us for this Scenario 3, as he is visiting with IWP personnel in Italy, but he has asked me to share with you that IWP is making good use of the recommendations that you folks hand over at the end of each seminary.

Again, many thanks to each and every one of you. See you all next week for the next Scenario 4, one represented by highly extreme conditions.

Chapter 8:
Scenario 4: Nuclear Warfare and Death of the Human Genome

*"**Hiroshima** has become a metaphor not just for **nuclear war** but for war and destruction and violence toward civilians. It's not just the idea we should not use nuclear arms. We should not start another war because it's madness."*
-- *Max von Sydow*
Read more at:
https://www.brainyquote.com/quotes/keywords/nuclear_war.html

*"We must **eliminate all nuclear weapons** in order to eliminate the grave risk they pose to our world. This will require persistent efforts by all countries and peoples. **A nuclear war** would affect everyone, and all have a stake in preventing this nightmare."*

--Please, please come in and have a seat. There is still plenty of room and chairs in this conference room. We should be able to begin our *Seminar 4* in twenty minutes. –One of the students in the *Department of Systems and Industrial Engineering (SIE)*, University of Arizona (UofA), Tucson, Arizona, USA, was saying to the people arriving to hear the presentation in the War Scenario Series.

Kathy Thompson

Half an hour later, everyone is seated, and waiting.

--Good afternoon, my name is *Kathy Thompson*, and would like to welcome all of you here today, at this conference room, to the presentation of *Scenario 4* with title:

"Nuclear Warfare and Death of the Human Genome"

Yes, we all know, it sounds very apocalyptic, destructive, and sad, but that is precisely what we would like to get into this afternoon, with the contents prepared by one of the research teams of the International Organization for World Peace (IWP) based here, at the University of Arizona. But before we begin I would like to remind you that we encourage questions and comments, during the presentation and afterwards, on the one-page survey sheet available on top of your desks.

One hand is up in the air in the audience.

--Welcome to our seminar! Yes, please, what is your question?

--Thank you. Will you be able to present statistics on the total number of nuclear warheads available in the world today, and which countries have the most warheads?

--Yes, we have that information in this presentation, and we like to share it with you, but first let us take a look at a textual description of a "nuclear holocaust" as it appears in our first PowerPoint slide:

(PowerPoint slide No. 1):

*"A **nuclear holocaust** or **nuclear apocalypse** is a scenario involving widespread destruction and radioactive fallout causing the collapse of civilization, through the use of nuclear weapons. Under such a scenario, some of the Earth is made uninhabitable by nuclear warfare in future world wars.*

*Besides the obvious direct destruction of cities by **nuclear blasts**, the potential aftermath of a nuclear war could involve **firestorms, a nuclear winter, widespread radiation sickness from fallout, and/or the temporary loss of much modern technology due to electromagnetic pulses**. Some scientists have speculated that a thermonuclear war could result in **the end of modern civilization on Earth,** in part due to a long-lasting nuclear winter. In one model, temperatures following a full thermonuclear war fall for several years by 7 to 8 degrees Celsius on average. The accuracy of such models is often the subject of partisan dispute.*

*Early Cold War-era studies suggested that billions of humans would nonetheless survive the immediate effects of nuclear blasts and radiation following a global thermonuclear war. Some scholars argue that nuclear war could indirectly contribute to **human extinction via secondary effects, including environmental consequences, societal breakdown, and economic collapse**. Additionally, it has been argued that even a relatively small-scale nuclear exchange between India and Pakistan, for example, involving 100 Hiroshima yield weapons, could cause a nuclear winter and **kill more than 1,000 Million people.** "*[1]

--There we are. We are talking about thousands of Millions of people dying in such a nuclear war, depending on the number of countries involved in such catastrophe. It would be the end of civilization and humanity as we know it today, with millions of people still alive left on the surface of our planet, but with radioactivity in their bodies from radioactive fallout, the end of agriculture, and with crime mafia networks along roads and cities trying to steal remaining food supplies. More Millions of people would die over the following months all over the planet. OK, will be getting deeper into the consequences of such a nuclear holocaust, but let us now take a look at the list of countries which have nuclear warheads.

Country Members of the NPT

There are 5 countries who are members of the *NPT*, the *Treaty of Non-Proliferation of Nuclear Weapons,* an international treaty intended to prevent the spread of nuclear weapons and technology, and to achieve nuclear disarmament, which was initiated in 1968, namely the United States, Russia, the United Kingdom, France, and China, as shown on *Figure 1*. Other countries included, but not belonging to the NPT, are India, Pakistan, and North Korea.

The United States was the first country to develop nuclear warheads, carried its first test in July 16, 1945, and went on to use nuclear warheads to bomb and devastate the Japanese cities of Hiroshima and Nagasaki during World War II. Next, during the Cold War that followed, the US built some 70,000 nuclear warheads.

The *United Kingdom* (UK) went on to test its first nuclear weapon in 1952, working with refugee scientists who had left other countries in the European continent following the end of World War II. In fact, the UK, the United States, and Canada collaborated in the Manhattan Project in the development and testing of nuclear warheads. The UK deployed its nuclear weapons on bomber aircraft and on ballistic missile submarines (SSBNs).

By contrast, *France* conducted its own research and testing of nuclear technology on its own, motivated by the *Suez Canal Crisis* on the 1960s. After the Cold War France went on to disarm its nuclear capability by 175 warheads, reducing its arsenal, and deploying some 300 warheads of a dual system of submarine-launched ballistic missiles (SLBMs) and medium-range air-to-surface missiles.

Country	Number of Warheads (Active/Total)	Date of first test	CTBT Status	Delivery Method
Russia	1,790 /7,300	29 August 1949	Ratifier	Nuclear triad
United States	1,750 /6,970	16 July 1945	Signatory	Nuclear triad
United Kingdom	150 /215	3 October 1952	Ratifier	Sea-Based
France	290 /300	13 February 1960	Ratifier	Sea and air-based
China	n.a. /260	16 October 1964	Signatory	Nuclear triad
India	n.a. /120	18-may-74	Non-signatory	Nuclear triad
Pakistan	n.a. /130	28-may-98	Non-Signatory	land and air-based
North Korea	n.a. /10	9 October 2006	Non-Signatory	land and sea-based
Israel	n.a. /400	1960-1979	Signatory	Nuclear triad

Figure 1 . List of Nations with Nuclear Warhead Weapons. [2]

Giant **China**, mainland, developed and first tested nuclear warheads in 1964 at the Lop Nur test site, as a deterrent against both the United States and Russia powers. It also tested its first "hydrogen bomb" in 1967, a very short time of 3 years after its first test. It is believed to have arsenal of 400 nuclear warheads today.

Among the countries which do not belong to the NPT is **India**, which went on to test and detonate its first nuclear warhead in 1974, calling it a "peaceful nuclear explosive." In 2005 US President *George W. Bush* and Indian Prime Minister *Manmohan Singh* announced plans to reach an indo-USA nuclear agreement. Today India still maintains an arsenal of 120 nuclear warheads.

Pakistan is not a member of the NPT organization either, but frequently it has been at odds with India. As such, Pakistan has been developing and testing nuclear warheads since the late 1970s with equipment and materials supplied by Western powers. It is estimated that this country maintains a stockpile of 130 nuclear warheads.

--Question, please. I would never have imagined a country like **Pakistan** having nuclear weapons. How did Pakistan get to develop such an expensive program of nuclear weapons?

--Good question, and the answer is that many factors entered into such a situation. Mainly, Pakistan and India were already having their differences and guerrilla conflicts; India already had nuclear weapons, and government officials in Pakistan did not want to be left behind. So, here is text with additional detail:

> *"Pakistan is not a party to the Nuclear Non-Proliferation Treaty. Pakistan **covertly** developed nuclear weapons over decades, beginning in the late 1970s. Pakistan first delved into nuclear power after the establishment of its first nuclear power plant near Karachi with equipment and materials **supplied mainly by western nations** in the early 1970s. Pakistani President Zulfiqar Ali Bhutto promised in 1971 that if India could build nuclear weapons then Pakistan would too, according to him: "**We will develop Nuclear stockpiles, even if we have to eat grass.**"*

> *It is believed that Pakistan has possessed nuclear weapons since the mid-1980s. The United States continued to certify that Pakistan did not possess such weapons until 1990, when sanctions were imposed under the Pressler Amendment, requiring a cutoff of U.S. economic and military assistance to Pakistan. In 1998, Pakistan conducted its first six nuclear tests at the Ras Koh Hills in*

response to the five tests conducted by India a few weeks before.

*In 2004, the Pakistani metallurgist Abdul Qadeer Khan, a key figure in Pakistan's nuclear weapons program, **confessed** to heading an international black market ring involved in selling nuclear weapons technology. In particular, Khan had been selling gas centrifuge technology to North Korea, Iran, and Libya. Khan denied complicity by the Pakistani government or Army, but this has been called into question by journalists and IAEA officials, and was later contradicted by statements from Khan himself."* [1]

--There you have it. Pakistan had the money, and a number of Western states went for that money selling weapon technology to Pakistan.

And then, there is **North Korea**, originally a member of the NPT treaty until it announced its withdrawal in 2003, after the USA accused it of having a secret uranium enrichment program. It conducted its first nuclear testing in 2006 claiming intimidation by the United States. Ever since, North Korea has continued in its development and testing of nuclear warheads, extending such activity to 2017, in the middle of a dramatic "Cold War" crisis with the United States.

Israel? We almost forgot about Israel! Yes, this country also has nuclear weapons, although it has not come forward to actually confirm so. *"Israel is not a party to the NPT, and purposely it engages in strategic ambiguity"*, is said in the global community. According to the Federation of American Scientists, Israel very likely has around 75-400 nuclear warheads for delivery in Jericho II medium-range ballistic missiles, of which 30 are gravity bombs to be delivered by aircraft.

Nuclear Warheads in the Planet

More nuclear warheads out there in the planet? Yes. Under the **North Atlantic Treaty Organization (NATO)**, the **United States** shares some of its own nuclear arsenal with several countries, including Belgium, Germany, Italy, Netherlands, and Turkey, as shown on *Figure 2*. Both USA forces and local armed forces join to maintain this capability against the possibility of a Russian attack. Belgium, for example, stockpiles 10-20 nuclear warheads at its air base in Kleine Brogel under the management of the 52nd Fighter Wing.[2]

Country	Air Base	Custodian	Warheads
Belgium	Kleine Brogel	52nd Fighter Wing	10 --20
Germany	Büchel	52nd Fighter Wing	20
Italy	Ghedi Torre	52nd Fighter	10--20
	Aviano	31st Fighter Wing	50
Netherlands	Volkel	52nd Fighter Wing	22
Turkey	Incirlik	39th Air Base Wing	60--70

Figure 2 . USA Nuclear Warheads in host Countires. [2]

Likelihood of a Nuclear Scenario

Another hand is up in the air amid the audience.

--Obviously, this is a very complex and extreme scenario, annihilation of our civilization through the destructive power of nuclear weapons. Could you give us an idea of *the likelihood of such scenario,* and a list of the main elements in it?

--Yes, I can share with you *Figure 7* in the next PowerPoint slide which shows critical elements in Scenario 4, such as Main Factors, Time Frame, some potential nuclear warfare among countries, and a list of "close calls" which we as a civilization have had in the last 60-70 years, during the so-called "Cold War."

--*Likelihood of such Scenario 4?* That is, the probability of occurrence of such scenario? A difficult question to answer, but perhaps not so difficult, really, especially after we realize that we have stockpiled thousands of nuclear warheads by a long list of countries, and after realizing that we have so many conflicts among these countries in areas of "national security", political ideologies, religious differences, and military objectives. Therefore, I personally would say there is *a 50%-60% probability of a nuclear war sometime in the next 15-20 years.* My personal opinion only. It is up to the entire IWP teams of experts to gather all their thoughts and evidence and give their verdict. Again, it does not have to be an "all out" war among many countries and, instead, it could be a war between two countries and the use of 4-6 nuclear warheads only. An example of such a war could be one between the USA and North Korea, the first being a large country, and the second one being a small country, both with very different political hierarchies in them. A miscalculation on either side could cause a first launch of nuclear missiles by one country followed by a retaliatory launch of missiles by the other side. Unfortunately, "intelligence" would most likely not be a determining factor and, instead, stupidity, irresponsibility, and fear would trigger such a holocaustic event.

And now I am going to ask my partner for this afternoon, *Xavier Elurmendi,* to continue with the second half of this presentation. Xabier, could you please join us?

Xabier Elurmendi

Thank you, Kathy, I am very pleased to be here to continue with the second half of our presentation. As Kathy already mentioned, my name is *Xabier Elurmendi*, and I am part of the IWP team presenting this series of extreme global scenarios.

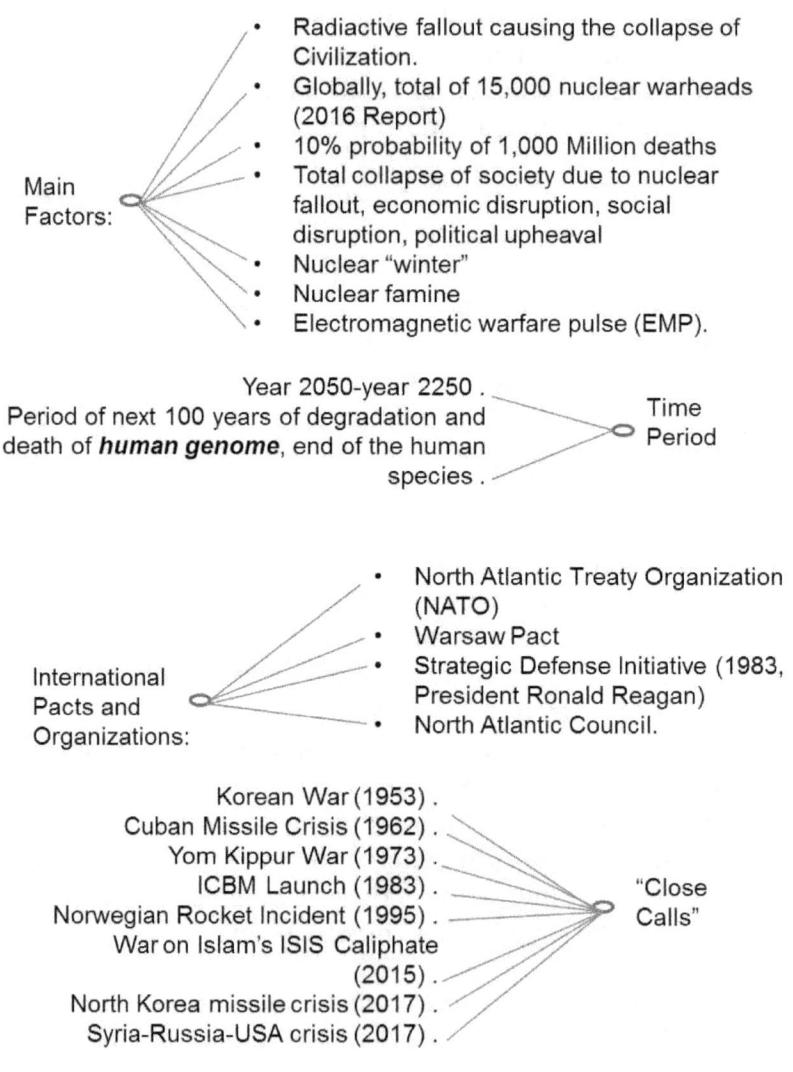

Main Factors:
- Radiactive fallout causing the collapse of Civilization.
- Globally, total of 15,000 nuclear warheads (2016 Report)
- 10% probability of 1,000 Million deaths
- Total collapse of society due to nuclear fallout, economic disruption, social disruption, political upheaval
- Nuclear "winter"
- Nuclear famine
- Electromagnetic warfare pulse (EMP).

Year 2050-year 2250. Period of next 100 years of degradation and death of **human genome**, end of the human species. — Time Period

International Pacts and Organizations:
- North Atlantic Treaty Organization (NATO)
- Warsaw Pact
- Strategic Defense Initiative (1983, President Ronald Reagan)
- North Atlantic Council.

Korean War (1953).
Cuban Missile Crisis (1962).
Yom Kippur War (1973).
ICBM Launch (1983).
Norwegian Rocket Incident (1995).
War on Islam's ISIS Caliphate (2015).
North Korea missile crisis (2017).
Syria-Russia-USA crisis (2017). — "Close Calls"

Figure 3. **Scenario 4**, Nuclear Warfare and Death of the Human Genome (Part 1 of 2)

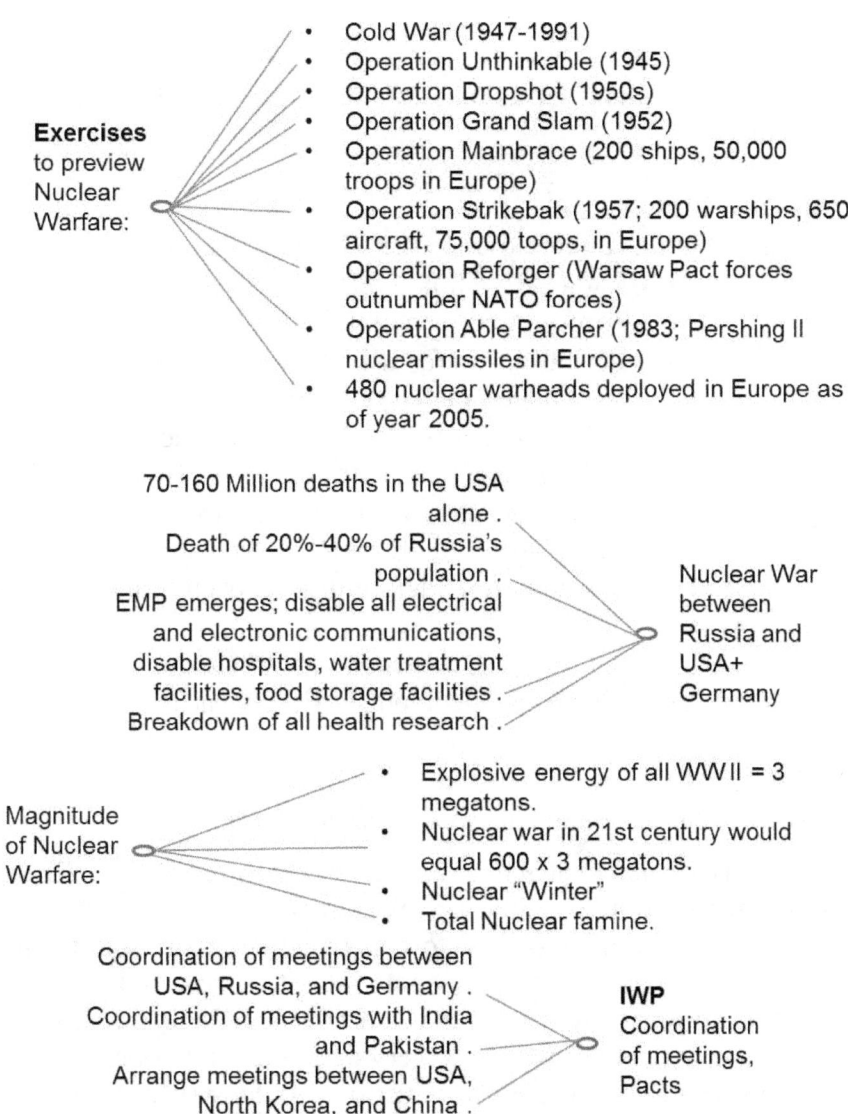

Exercises to preview Nuclear Warfare:

- Cold War (1947-1991)
- Operation Unthinkable (1945)
- Operation Dropshot (1950s)
- Operation Grand Slam (1952)
- Operation Mainbrace (200 ships, 50,000 troops in Europe)
- Operation Strikebak (1957; 200 warships, 650 aircraft, 75,000 toops, in Europe)
- Operation Reforger (Warsaw Pact forces outnumber NATO forces)
- Operation Able Parcher (1983; Pershing II nuclear missiles in Europe)
- 480 nuclear warheads deployed in Europe as of year 2005.

70-160 Million deaths in the USA alone .
Death of 20%-40% of Russia's population .
EMP emerges; disable all electrical and electronic communications, disable hospitals, water treatment facilities, food storage facilities .
Breakdown of all health research .

Nuclear War between Russia and USA+ Germany

Magnitude of Nuclear Warfare:

- Explosive energy of all WWII = 3 megatons.
- Nuclear war in 21st century would equal 600 x 3 megatons.
- Nuclear "Winter"
- Total Nuclear famine.

Coordination of meetings between USA, Russia, and Germany .
Coordination of meetings with India and Pakistan .
Arrange meetings between USA, North Korea, and China .

IWP Coordination of meetings, Pacts

Figure 4. **Scenario 4**, Nuclear Warfare and Death of the Human Genome (Part 2 of 2)

To continue, then, let us take a look at the main factors in this *Scenario 4*:

Main Factors

Included in this Scenario 4 are the following factors:

- Radioactive fallout causing the collapse of a large portion of human civilization.
- Globally, there are 15,000 nuclear warheads stockpiled in a list of countries.
- 10% probability of 1,000 Million deaths on a first and limited nuclear strike.
- Total collapse of society due to nuclear fallout, economic disruption, social disruption, destruction of agricultural capacity, and political upheaval.
- A following *"nuclear winter"*, and
- Electromagnetic pulse (EMP) warfare.

Nuclear Winter

--And what is the meaning of *"nuclear winter"*?

--Yes. By *"nuclear winter"* we mean the severe *"global climatic cooling effect"* to occur after a nuclear strike occurs, which would initiate a large number of firestorms, which in turn would inject large amounts of ashes and soot into the stratosphere, thus preventing the sun rays to enter the atmosphere. For months, and possibly years, the days would be very dark, the temperatures would reach low levels, agriculture production would stop and in many parts of the planet the flora and fauna would disappear.

--Thank you.

War Exercises

--In our next PowerPoint slide we are going to long list of "war exercises" carried out by several countries in order to anticipate the "real nuclear war" which might occur:

"War Exercises"[3]:

Operation Dropshot. It was carried out in the 1950s by the ***USA*** as a contingency plan for a nuclear and conventional war with the ***Soviet Union***. This exercise envisioned "the use of 300 nuclear warheads, and 29,000 high explosive bombs on 200 targets in 100 towns and cities in the Soviet Union to wipe out 85% of its industrial capability."

Operation Unthinkable. Carried out in April-May 1945, the ***British*** Armed Forces developed this operation as a scenario for World War III. "Prime Minister Winston Churchill was concerned that, with the enormous size of Soviet Forces deployed in Europe at the end of World War II and the unreliability of the Soviet leader Joseph Stalin, there would be a serious threat to Western Europe."

Exercise Mainbrace. "It brought together 200 ships and over 50,000 personnel to practice the defence of Denmark and Norway from Russian attack in 1952. It was a first major NATO exercise, jointly commanded by Supreme Allied Commander Atlantic Admiral Lynde D. McCormick, USN, and Supreme Allied Commander Europe General Mathew B. Ridgeway, US Army, during the Fall of 1952."

Exercise Grand Slam and Longstep. "Naval exercises held in the Mediterranean Sea during 1952 to practice dislodging an enemy amphibious attack. It involved over 170 warships and 700 aircraft." The overall commander, Admiral Carney, summarized the exercise saying: "We have demonstrated that the senior commanders of all four powers can successfully take charge or a mixed task force and handle it effectively as a working unit."

Operation Strikeback. "This was a major North Atlantic Treaty Organization (NATO) naval exercise held in 1957, simulating a response to an all-out Soviet attack on NATO. The exercise involved 200 warships, 650 aircraft, and 75,000 personnel from the United States Navy, the United Kingdom's Royal Navy, the Royal Canadian Navy, the French Navy, the Royal Netherlands Navy, and the Royal Norwegian Navy."

Operation Reforger. "An annual exercise conducted during the Cold War (1947-1991) by NATO. The objective was to show that NATO had the ability to quickly deploy forces to West Germany in the event of a conflict with the Warsaw Pact. Most of the support would come across the Atlantic Sea, from USA and Canada."

Exercise Able Archer. "A five-day NATO war exercise which started on 7 November 1983, spanned Western Europe, and was completed on 11 November, 4 days later; it simulated a period of conflict escalation, ending in a coordinated nuclear war. It took place in response to deteriorating relations between the USA and the Soviet Union, as a result of the anticipated arrival of strategic Pershing II nuclear missiles in Europe from the USA. The poor relations had been triggered by the Strategic Defense Initiative (SDI) proposed by the US President Ronald Reagan on 23 March 1983."

--Yes, Mr. Alurmendi, some of us still remember some of those war exercises, however, what do we know today about some events or sets of circumstances that got us very close to an actual nuclear warfare? Perhaps we can learn from those events and stay away from the auto-destruction of a nuclear war.

--I hear you, and there is indeed a list of historical *"close calls"*, events which got us all in the planet very close to an actual outbreak of a nuclear war. To that end, let us take a look at the next set of PowerPoint slides.

"Close Calls" [3]:

The Korean War (25 June 1950-27 July 1953). "A war between two factions trying to control the Korean peninsula: a Communist faction supported by China and the USSR, and a capitalist faction supported by the United States and the United Nations (UN). War correspondent Bill Downs wrote: *"Yes, Korea is the beginning of World War III. The brilliant landings at Inchon and the cooperative efforts of the USA and UN as allies have won us a victory in Korea. But this is only the first battle in a major international struggle which is now engulfing the Far East and the entire world."*

The Cuban Missile Crisis (15-28 October 1962). "A confrontation on the stationing of Soviet nuclear missiles in Cuba, in response to the failed Bay of Pigs invasion by the USA, is considered to have been the closest to a nuclear weapon exchange, which could have triggered World War III. The crisis peaked on 27 October of the same year when a U-2 plane was shot down over Cuba and another almost intercepted over Siberia."

The Yom Kippur War (6-25 October 1973). "Also known as the Ramadan War, began with Arab victories; Israeli forces successfully counterattacked. Next, tensions grew between the USA which supported Israel, and the Soviet Union which supported the Arab states. American and Soviet naval forces came close to firing on each other. That was another "close call" to a nuclear war. Tension eased with the ceasefire brought in under the agreement UNSC 339 of the United Nations."

NORAD (9 November 1979). "The USA made emergency retaliation preparations after the North American Aerospace Defense Command (NORAD) saw on computer screen what appeared to be indications of a full-scale Soviet attack. No attempt was made to use the ***"red telephone"*** hotline to clarify the situation with the USSR, and it was not until early-warning radar systems confirmed no such lunch had taken place. A Senator inside the NORAD facility at the time described an atmosphere of ***absolute panic***."

False Alarm (26 September 1983). "A false alarm occurred on the Soviet nuclear early warning system, showing the lunch of American Minutemen ICBM from bases in the USA. A retaliatory strike was prevented by *Stanislav Petrov*, an officer of the Soviet Air Defense Forces, who realized the system had malfunctioned."

Operation Able Archer 83 (2-11 November 1983). "During the Operation Able Archer, a 10-day NATO exercise simulating a period of conflict escalation which would culminate in a nuclear strike, some member of the Soviet Politburo and armed forces treated the events as a genuine first strike. In response, the Soviets prepared for a coordinated counter-attack by readying nuclear forces. Fortunately, the Soviet preparations for retaliation ceased upon completion of the Able Archer exercises."

Norwegian Rocket incident (25 January 1995). "This nearly fatal incident occurred when the Russian Federation Olenegorsk early warning station accidentally mistook the radar signature from a Black Brant XII research rocket, being jointly launched by Norwegian and USA scientists, as appearing to be the launch of a Trident SLBM missile. In response President Boris Yeltsin was summoned and the Cheget nuclear briefcase was activated for the first time. The Soviet high command soon realized that the rocket was not entering Russian airspace. The Russian radar technicians had not been informed ahead of time of the Norwegian war exercise."

North Korea Incident (5-18 April 2017).
"*April 5:* U.S. Secretary of States reacts to N.K. latest missile test with a brief a cryptic statement: "*North Korea launched yet another intermediate range ballistic missile. The United States has spoken enough about North Korea. We have no further comment.*" In a phone call to Prime Minister Shinzo Abe, U.S. president ***Trump*** promises to boost US military capabilities after Pyongyang fired ballistic missile.
April 6: Mongolia deregisters more North Korean vessels, following UN Security Council Resolution 2321.
April 6: U.S. bombs Syria to punish the regime's use of chemical weapons. It is also seen as the Trump administration signaling N.K. its willingness to use military force to compel N.K. to stop its development of nuclear bombing capabilities.
April 7: China and U.S. leaders Trump and Xi meet. Trump seeks Xi's cooperation in dealing with N.K., but states he is prepared to act alone. No specific commitments resulted from this meeting.

April 8: The U.S. announces the rerouting of the ***Carl Vinson Strike Group*** (consisting of an aircraft carrier and other warships) from its original planned route from Singapore to Australia, to the West Pacific, near the Korean Peninsula. This is in response to N.K.'s recent nuclear and missile tests, which the U.S. calls the *"The number one threat in the region"*, and discourage further tests. There are also recent indications from North Korea that it may be about to test an intercontinental missile.

April 12: China orders its military to be on nationwide alert and ready to move, in areas North Korea border, as tensions escalate on the peninsula.

April 12: China's leader Xi calls U.S. President Trump to advocate for a *"**peaceful resolution**"* in tensions with N.K.

April 13: Trump warns N.K. to back down from a soon-expected nuclear test. Trump's remark are taken as a threat of military action against the North.

April 14: The Chinese government is reported to be strengthening its diplomatic efforts to diffuse tensions between N.K. and the U.S. China's foreign minister Wang Yi states that *"The United States and South Korea and North Korea are engaging in tit for tat, with swords drawn and bows bent, and there have been storm clouds gathering,"*

April 18: It is revealed that when the Carl Vinson aircraft group was announced on April 8 to be heading to the Korean peninsula as a deterrent to N.K., due to a "glitch-ridden sequence of events" it was actually heading in the opposite direction. Finally, it did change course and start heading there, with an arrival expected a week later. A "close call" to a nuclear war had come to a peaceful end. Thus far. [3]

Destructive Capability of a Nuclear Warhead

--Yes, those were very "close calls", as those events came close to initiating a nuclear war, each one of them. But, what is the magnitude of the destructive capability of each nuclear warhead? Are they by now 5-10 times more destructive than the nuclear warhead thrown down at Hiroshima, for example.

--Much more, really. The nuclear warhead thrown down at **Hiroshima** during World War II had a destructive capability of 3 megatons, and it killed 20,000 soldiers along with another 126,000 civilians. An enormous tragedy. A nuclear warhead today can be as 6,000 times more destructive than the nuclear warhead which destroyed Hiroshima, so we are talking about Millions of people who would be killed with a nuclear warhead in today's stockpiles.

Potential Nuclear Wars in this Century

--Yes, Mr. Elurmendi, such devastating powers and such number or atomic warheads stockpiled today represent a great danger to humanity, no question about it. So, if I may, I would like to ask you about the countries which could initiate a nuclear warhead, is it just Russia and USA?

--No, not really, according to our IWP team in consultation with other international organizations there could be dozens if incidents which could trigger nuclear wars, wars which could begin with 4-6 nuclear warheads followed by retaliation attacks using dozens of nuclear warheads and more. Which countries? Well, we could have several countries entering such a devastating conflict for humanity, including:

- **Russia** and **USA** attacking each other, each country with its own set of allied countries, of course, with 70-160 Million deaths in the USA, and 20%-40% of the Russian population also dying.

- **Israel** and some **Arab countries** attacking each other, again the death toll being counted in the Millions of people dead in such event.

- **India** and **Pakistan** attacking each other, each one counting with populations of 1,326 Million and 197 Million, respectively. Again a massacre of terrible proportions with Millions of people dead.

- **USA** and **North Korea** attacking each other in this 21st Century; one being a very large country and the other one a very small countries. Again, lack of political maturity and military experience could trigger such a nuclear holocaust.

- *Iran* and *Israel* attacking each other due to political and ideological differences.

- *China* and *USA* attacking each other, with populations of 1,373 Million and 325 Million, respectively.

Now these are hypothetical nuclear warfare encounters, only, and the probability of occurrence of any of these encounters is a most difficult task. – Commented Xabier Elurmendi, responding to the last question – But, on a personal level, I would like to add that such probability is increased as the level of military and political experience of the leaders of the opposing countries reaches low levels, I would say.

Another hand is seen up in the air in the audience.

Death of the Human Genome

--What can the international IWP organization do to try to prevent such nuclear conflicts, and why in the title of this Seminar 4 we have the words "*Death of the Human Genome*"? Thank you.

--Yes, thank you for your question. Well, what we are saying is that such nuclear wars cause Millions of deaths across the cities and towns of the countries involved in such conflict, for one. What we are also saying is that such catastrophic events and circumstances would be aggravated by the emergence of national and international criminal organizations in search of food and spaces nearly free of radiation. As we have said before, in earlier seminars, the agricultural lands in the aftermath of a nuclear holocaust would be highly contaminated with radioactive minerals and other materials, the "*nuclear winter*" would hardly allow sunlight to penetrate the dark skies. As such we would be contemplating the "*death of the human genome*", a slow but certain death of the gene structure in our bodies, for sure.

--*What can the IWP organization do to prevent such catastrophic nuclear warfares*? Well, what IWP is doing across its many centers in many countries is helping us become aware of the extreme dangers of such conflicts with conferences and seminars like this Seminar 4. Other international conferences are being held in other countries to become aware of the large stockpile of nuclear warhead in the world today, the danger of miscalculations among political and military leaders, etc.

Again, I would like to thank everyone here today for your questions and comments on behalf of Dr. Eugene Finley, Kathy, and myself and, yes, please don´t forget to fill in the one-page questionnaire with your recommendations to IWP.

Thank you all.

✳✳✳

Chapter 9:
Scenario 5: Diversity of Religions and Beliefs in the World

*"I'm **sickened by all religions. Religion has divided people.** I don't think there's any difference between the pope wearing a large hat and parading around with a smoking purse and an African painting his face white and praying to a rock."*

-- **Howard Stern**
Read more at:
https://www.brainyquote.com/quotes/keywords/religions.html

*"We are a country where people of all backgrounds, all nations of origin, all languages, **all religions**, all races, can make a home. America was built by immigrants."*

-- **Hillary Clinton**
Read more at:
https://www.brainyquote.com/quotes/keyword

153

s/religions.htmlww.brainyquote.com/quotes/k eywords/overpopulation.html

Can a world such as ours, with a great diversity of religions and beliefs come to an understanding, or is such diversity a main obstacle to world peace and development in the future? This is one of the main questions we address in this chapter. Towards that end we will listen to the contents of *Scenario 5* as presented by *Kathy Thompson* and *Xabier Elurmendi*, member of the research team at the ***International Organization for World Peace (IWP)***

--Please, come in, there are plenty of seat in this conference room, make yourselves comfortable. My name is ***Xabier Elurmendi***, a member of the ***International World Organization for Peace (IWP),*** an organization which many of you are already familiar with. Great! Well, this afternoon we are going to look at still another world scenario where the main topic is ***religions*** and other beliefs as can be ***Atheism***, that is, the absence of a religion or belief in a supernatural entity. In fact, the title of our Seminar 5 is:

"Diversity of Religions and Beliefs in the World"

and thus we are going to try, consider, and also debate whether our world would be better off with a large diversity of religions and beliefs or, on the other hand, whether we would be better off without any religions, Ok? To get started, we are going to take a condensed look at this Scenario 5 in ***Figure*** 1, in the next two PowerPoint slides, if you will, please.

Now, I must also add that this Scenario 5 is to be found within the category of ***"negotiate-and-live"***, as we mentioned in earlier seminars. So, to get started, are there any questions on this subject?

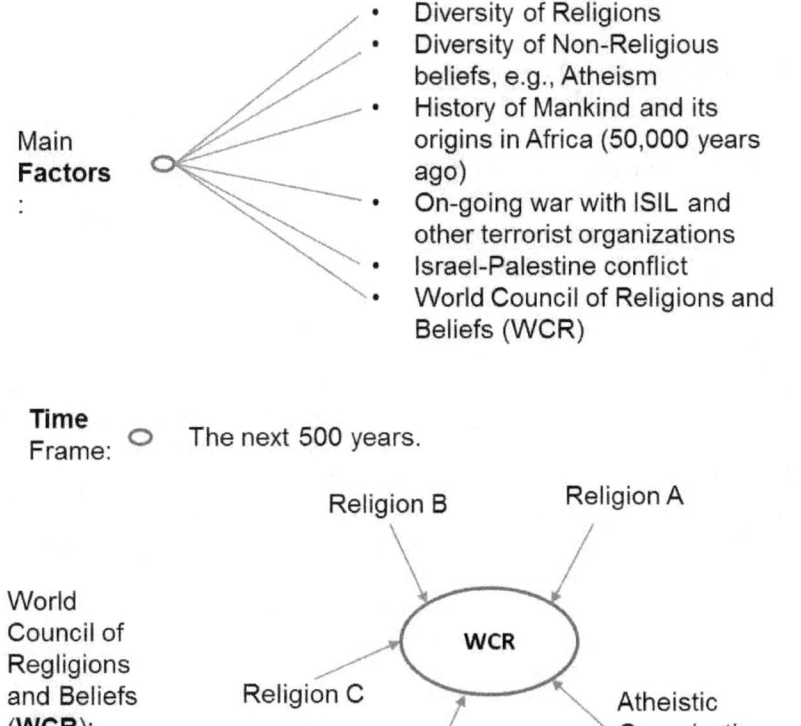

Main **Factors** :
- Diversity of Religions
- Diversity of Non-Religious beliefs, e.g., Atheism
- History of Mankind and its origins in Africa (50,000 years ago)
- On-going war with ISIL and other terrorist organizations
- Israel-Palestine conflict
- World Council of Religions and Beliefs (WCR)

Time Frame: The next 500 years.

World Council of Regligions and Beliefs (WCR):

Religion B   Religion A

WCR

Religion C

Atheistic Organizations

Religion N

CWR Principles:
- Every person has the right to believe or not to believe in a Religion.
- Religious leaders of a particular Religion are not to preach to members of other Religions.
- Religious leaders are not to interfere in any way with political activities, movements, or programs.
- All ethnic groups in the Planet are equal, and have the right to be heard by the WCR.

Figure 1. Scenario 5, Diversity of Religions and Beliefs in the World. (Part 1 of 2)

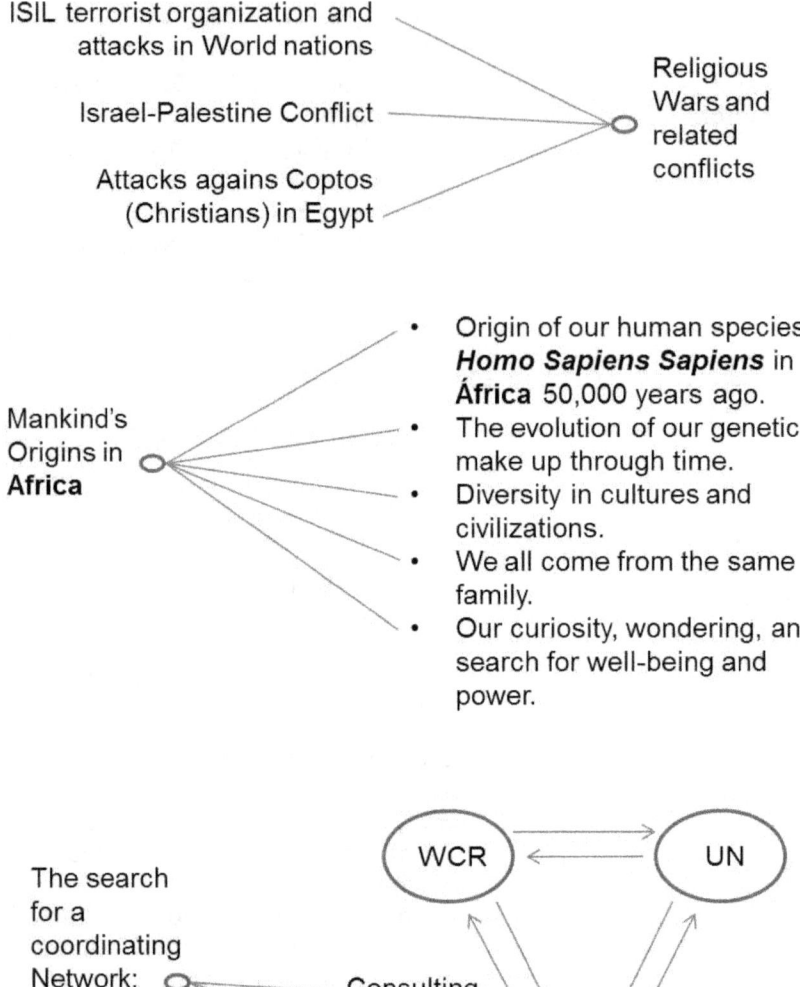

ISIL terrorist organization and attacks in World nations

Israel-Palestine Conflict

Attacks agains Coptos (Christians) in Egypt

Religious Wars and related conflicts

Mankind's Origins in **Africa**

- Origin of our human species, **Homo Sapiens Sapiens** in **África** 50,000 years ago.
- The evolution of our genetic make up through time.
- Diversity in cultures and civilizations.
- We all come from the same family.
- Our curiosity, wondering, and search for well-being and power.

The search for a coordinating Network:

WCR

UN

Consulting, Decision-Making, and Agreements

IWP

Figure 1. Scenario 5, Diversity of Religions and Beliefs in the World. (Part 2 of 2)

A hand is raised in the audience.

--Yes, *Mr. Peterson*, please go ahead with your question. As you can see, by now we are beginning to know each other by name.

--Thank you, *Xabier*. I wonder if you could comment because by now many of us in our world believe that religions have played a major role in wars for hundreds and thousands of years. Would it be better to consider a world where all religions are abolished, and where all races become knowledgeable of our common origin in Africa some 50,000 years ago?

--Well, that is a very good question, and I assure you Mr. Peterson that our research team has considered such a scenario. One major situation to consider is that today we have hundreds of religions in our planet, the result of thousands of years of social evolution, and that as you may realize it would take that much time to try convince ourselves that a world without religions is the answer that we are looking for and, unfortunately, we may not have that much time left as a species. So, I am going to ask my partner in this Seminar Series, *Miss Kathy Thompson*, to help us project the PowerPoint slides to so we can see some of the major differences and similarities among the major religions. Kathy, if you will please.

(PowerPoint slide 1):

Christianity

An *Abrahamic monotheistic religion* based   on the life and teachings of *Jesus Christ*, who serves as the focal point for the religion. It is the world's largest religion, with over *2,400 Million followers,* or 33% of the global population, known as Christians. Christians believe that Jesus is the Son of God and the savior of humanity whose coming as the Messiah (the Christ) was prophesied in the Old Testament.

In the 2nd century, Christianity was criticized by the Jews on various grounds, e.g. that the prophecies of the Hebrew Bible could not have been fulfilled by Jesus, given that he did not have a

157

successful life. Additionally a sacrifice to remove sins in advance, for everyone or as a human being, did not fit to the Jewish sacrifice ritual; furthermore God is said to judge people on their deeds instead of their beliefs.

Their heaven? A place "out there", "somewhere", where all the believers live happily ever after in the community with God, the Virgin Mary, and all the saints.

(PowerPoint slide 2):

Judaism

Judaism is another ancient monotheistic Abrahamic religion, with the Torah as its foundational text (part of the larger text known as the Tanakh or Hebrew Bible), and supplemental oral tradition represented by later texts such as the Midrash and the Talmud. Judaism is considered by religious Jews to be the expression of the covenantal relationship that God established with the Children of Israel. With between 14.5 and 17.4 million adherents worldwide, Judaism is the tenth-largest religion in the world.

The total number of Jews worldwide is difficult to assess because the definition of "who is a Jew" is problematic; not all Jews identify themselves as Jewish, and some who identify as Jewish are not considered so by other Jews. According to the *Jewish Year Book* (1901), the global Jewish population in 1900 was around 11 million. The latest available data is from the World Jewish Population Survey of 2002 and the Jewish Year Calendar (2005). In 2002, according to the **Jewish Population Survey**, there were 13.3 million Jews around the world. The Jewish Year Calendar cites 14.6 million.

While the concept of **Heaven** is much discussed within the Christian and Islamic religions, the Jewish concept of the afterlife, sometimes known as *olam haba*, the World-to-come, is not discussed so often. The Torah has little to say on the subject of survival after death, but by the time of the rabbis two ideas had made inroads among the Jews: one, which is probably derived from Greek thought, is that of the immortal soul which returns to its creator after death; the other, which is thought to be of Persian origin, is that of resurrection of the dead.

(PowerPoint slide 3):

Islam

Another Abrahamic monotheistic religion which professes that there is only one and incomparable God (Allah) and that Muhammad is the last messenger of God. It is the world's second-largest religion and the fastest-growing major religion in the world, with over 1,700 Million followers or 23% of the global population, known as Muslims. Islam teaches that God is merciful, all-powerful, and unique; and He has guided mankind through revealed scriptures, natural signs, and a line of prophets sealed by Muhammad. The primary scriptures of Islam are the Quran, viewed by Muslims as the verbatim word of God, and the teachings and normative example (called the *Sunnah*, composed of accounts called *hadith*) of Muhammad (c. 570–8 June 632 CE). The cities of Mecca, Medina and Jerusalem are home to the three holiest sites in Islam.

Criticism. Objects of criticism include the morality of the life of Muhammad, the last law bearing prophet of Islam, both in his public and personal life, as seen in medieval Christian views on Muhammad. Issues relating to the authenticity and morality of the Quran, the Islamic holy book, are also discussed by critics. Other criticisms focus on the question of human rights in modern Islamic nations, and the treatment of women in Islamic law and practice. In wake of the recent multiculturalism trend, Islam's influence on the ability of Muslim immigrants in the West to assimilate has been criticized.

(PowerPoint slide 4):

Buddhism

Buddhism is an Indian religion and *dharma* that encompasses a variety of traditions, beliefs and spiritual practices largely based on teachings attributed to the **Buddha**. Buddhism originated in Ancient India sometime between the 6th and 4th centuries BCE (Before Christ), from where it spread through much of Asia, where after it declined in India during the middle ages. Two major extant branches of Buddhism are generally recognized by scholars: Theravada (Pali: "The School of the Elders") and Mahayana (Sanskrit: "The Great Vehicle"). Buddhism is the world's fourth-largest religion, with over *500 million followers* or 7% of the global population, known as Buddhists.

Buddhist schools vary on the exact nature of the path to liberation, the importance and canonicity of various teachings and scriptures, and especially their respective practices. Practices of Buddhism include taking refuge in the **Buddha**, the **Dharma** and the **Sangha**, study of scriptures, observance of moral precepts, renunciation of craving and attachment, the practice of meditation (including calm and insight), the cultivation of wisdom, loving-kindness and compassion, the Mahayana practice of bodhicitta and the Vajrayana practices of generation stage and completion stage.

In Buddhism there are several **Heavens**, all of which are still part of *samsara* (illusionary reality). Those who accumulate good karma may be reborn in one of them. However, their stay in Heaven is not eternal, as eventually they will use up their good karma and will undergo rebirth into another realm, as a human, animal or other being. Because Heaven is temporary and part of *samsara*, Buddhists focus more on escaping the cycle of rebirth and reaching enlightenment (**nirvana**). Nirvana is not a heaven but a mental state.

According to Buddhist cosmology the universe is impermanent and beings transmigrate through a number of existential "planes" in which this human world is only one "realm" or "path". These are traditionally envisioned as a vertical continuum with the Heavens existing above the human realm, and the realms of the animals, hungry ghosts and **Hell** beings existing beneath it. According to Jan Chozen Bays in her book, *Jizo: Guardian of Children, Travelers, and Other Voyagers*, the realm of the *asura* is a later refinement of the heavenly realm and was inserted between the human realm and the Heavens. One important Buddhist Heaven is the *Trāyastriṃśa*, which resembles Olympus of Greek mythology.

(PowerPoint slide 5):

Hinduism

Hinduism is a religion, or a way of life, found most notably in India and Nepal. Hinduism has been called the oldest religion in the world, and some practitioners and scholars refer to it as *Sanātana Dharma*, "the eternal law," or the "eternal way," beyond human origins. Scholars regard Hinduism as a fusion or synthesis of various Indian cultures and traditions, with diverse roots and no founder. This "Hindu synthesis" started to develop between 500 BCE and 300 CE[] following the Vedic period (1500 BCE to 500 BCE).

Although Hinduism contains a broad range of philosophies, it is linked by shared concepts, recognizable rituals, cosmology, shared textual resources, and pilgrimage to sacred sites. Hindu texts are classified into Shruti ("heard") and Smriti ("remembered"). These texts
discuss theology, philosophy, mythology, Vedic yajna, Yoga, agami crituals, and temple building, among other topics. Major scriptures include the Vedas and Upanishads, the Bhagavad Gita, and the Agamas. Sources of authority and eternal truths in its texts play an important role, but there is also a strong Hindu tradition of the questioning of this authority, to deepen the understanding of these truths and to further develop the tradition. Hinduism is the world's third largest religion, with over *1,000 Million followers* or 15% of the global population, known as Hindus. The majority of Hindus reside in India, Nepal, Mauritius, the Caribbean, and Bali in Indonesia.

Heaven. Attaining heaven is not the final pursuit in Hinduism as heaven itself is ephemeral and related to physical body. Only being tied by the bhoot-tatvas, heaven cannot be perfect either and is just another name for pleasurable and mundane material life. According to Hindu cosmology, above the earthly plane, are other planes: (1) Bhuva Loka, (2) Swarga Loka, meaning Good Kingdom, is the general name for heaven in Hinduism, a heavenly paradise of pleasure, where most of the Hindu Devatas (Deva) reside along with the king of Devas, Indra, and beatified mortals. Some other planes are Mahar Loka, Jana Loka, Tapa Loka and Satya Loka. Since heavenly abodes are also tied to the cycle of birth and death, any dweller of Heaven or **Hell** will again be recycled to a different plane and in a different form per the karma and "maya" i.e. the illusion of Samsara. This cycle is broken only by self-realization by the Jivatma. This self-realization is Moksha (Turiya, Kaivalya).

(PowerPoint slide 6):

Atheism

Atheism is, in the broadest sense, the absence of belief in the existence of deities. Less broadly, atheism is the rejection of belief that any deities exist. In an even narrower sense, atheism is specifically the position that there are no deities. Atheism is contrasted with theism, which, in its most general form, is the belief that at least one deity exists.

The etymological root for the word *atheism* originated before the 5th century BCE from the ancient Greek (*atheos*), meaning "without god(s)". In antiquity it had multiple uses as a pejorative term applied to those thought to reject the gods worshiped by the larger society, those who were forsaken by the gods or those who had no commitment to belief in the gods. The term denoted a social category created by orthodox religionists into which those who did not share their religious beliefs were placed. The actual term *atheism* emerged first in the 16th century. With the spread of freethought, skeptical inquiry, and subsequent increase in criticism of religion, application of the term narrowed in scope. The first individuals to identify themselves using the word *atheist* lived in the 18th century during the Age of Enlightenment. The French Revolution, noted for its "unprecedented atheism," witnessed the first major political movement in history to advocate for the supremacy of human reason.

Arguments for atheism range from the philosophical to social and historical approaches. Rationales for not believing in deities include arguments that there is a lack of empirical evidence, the problem of evil; the argument from inconsistent revelations, the rejection of concepts that cannot be falsified, and the argument from nonbelief. Although some atheists have adopted secular philosophies (e.g. secular humanism), there is no one ideology or set of behaviors to which all atheists adhere. Atheism is a more parsimonious position than theism and is the position in which everyone is born; therefore it has been argued that the burden of proof lies not on the atheist to disprove the existence of God but on the theist to provide a rationale for theism.

Since conceptions of atheism vary, accurate estimations of current numbers of atheists are difficult. Several comprehensive global polls on the subject have been conducted by Gallup International: their 2015 poll featured over 64,000 respondents and indicated that 11% were "convinced atheists" whereas an earlier 2012 poll found that 13% of respondents were "convinced atheists." An older survey by the British Broadcasting Corporation (BBC) in 2004 recorded atheists as comprising 8% of the world's population. Other older estimates have indicated that atheists comprise 2% of the world's population, while the irreligious add a further 12%. According to these polls, Europe and East Asia are the regions with the highest rates of atheism. In 2015, 61% of 1,350 Million people in China reported that they were atheists.

The International World Organization for Peace (IWP)

--Thank you, Kathy. As you all can see, religions are highly complex social models which we have developed over thousands of years, with contents and rules over the concepts of obedience, respect, fear, and salvation in an "after-life." We also would like to realize that in hundreds of our communities throughout the planet we have a few people with great economic means, while a great majority in the planet has the bare essentials to live, while aspiring at a better life in a "heaven" which is waiting for them provided they follow the rules of their particular religion, and their associated hierarchical power structures, of course. It is for these reasons that the IWP is proposing to reach an understanding, a cooperative structure, among the religions of the world, as suggested in Figure 1.

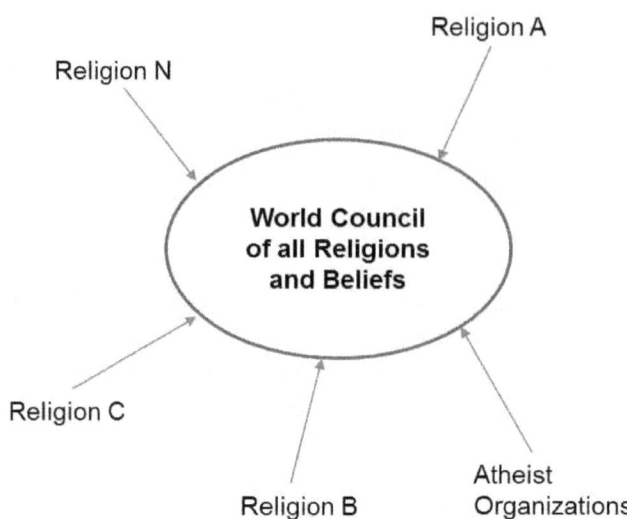

Figure 2. Proposal of a World Council of all Religions and Beliefs (WCR).

Another hand shows up in the audience.

--Well, it is a fine idea, but would it work? Also, what time frame are we talking about to create and to make work such *World Council of all Religions and Beliefs (WCR)*, and which would be the tenets or principles of such organization?

--Right, good questions. Our IWP organization is ready to propose such a Council within the next few months to several other international organizations but –and you are right – it would take decades for such Council to function and to become a true participant in international matters such as religious wars, food supply programs, reduction of numbers of nuclear warheads, and so forth. Regarding its tenets or principles, the next PowerPoint slide shows *a list of proposed principles*:

(PowerPoint slide 7):

Principle 1: Every person has the right to believe or not to believe in a Religion, and such right shall be respected by all members of the World Council of all Religions and Beliefs (WCR).

Principle 2: Religious leaders of a particular Religion are not to preach to members of other Religions.

Principle 3: Religious leaders are not to interfere in any way with political activities, movements, or programs; similarly, political leaders are not to interfere in any way with religious activities, movements, or programs.

Principle 4: All ethnic groups in the Planet are equal, and have the right to be heard by the WCR.

Principle 5: All WCR members adhere to and sponsor the concept of diversity in our planet with regards to thinking, programs, ways of life, and the pursuit of happiness.

--Again, Xabier, those principles sound great, but there are some facts that are not found among the books and tenets of religions of the world, such as the fact that all of mankind in the planet originated in Africa some 50,000-70,000 years ago, for example. Right now within most religions the belief is that humans were pro-created by deities, by supernatural powers, and that some peoples or communities have special relationships with those deities. What do we want to do about, or can do, about this situation?

--Yes, our IWP organization is aware of such knowledge, and that is why in the next PowerPoint slides we present information that we propose to share with our world communities regarding the origins of mankind.

(PowerPoint slide 8):

The Time Scale of the Evolution

A quick look at the timeline of evolution of the Earth, our own planet, and that of all the species, as shown on *Table 1*, might help us begin to understand the fragility of our own species in the planet.

Table 1. Timeline of Human Evolution	
Millions of Years Ago:	*Event:*
5000	The Earth is created (Big-bang theory).
4500	By then the surface of the Earth has cooled down, mostly.
4500-3800	**Hadean** period. Life begins at sea.
3800-2500	**Archean** period. Beginning of photosynthesis process.
2500- 600	**Proterozoic** period. Abundance of oxygen in the atmosphere.
540- 490	**Cambrian** period. "Pikaia", first "fish" with a spinal chord, the origin of all mammals
299- 200	95% of all species disappear.
200- 150	**Jurasic** period. Both "cold-blooded" and "warm-blooded" animals begin, first brids, first plants with flowers.
299- 200	95% de las especies desaparecen.
5-7	*Homo Ergaster,* common ancestro of all homo species and chimpancés.
0.3	*Neanderthals* our of Africa, into Europe and Asia (300,000 years ago).
0.04-0.6	*Homo Sapiens Sapiens* out of Africa (40,000-60,000 years ago). Families of this Group spread over the five continents. All human races in the planet today belong to this group.
Thousands of Years In the Future:	*Event:*
¿? - ¿?	Extinction of *Homo Sapiens Sapiens*.

How long ago? The Earth formed 5,000 million years ago, according to the "big-bang" theory. Yes, a very long time ago, it's nearly impossible to consider the magnitude of such long period of time. A big ball of fire rotating around the sun in our galaxy. Then, during the next 500 million years the Earth continued to cool down, the surface continued to cool down only, that is. Some time "shortly", 4,500-3,800 million years ago, during the Hadean period, *life began at sea*. Another 1,300 million years had to go by for the photosynthesis process to come into place, replacing carbon dioxide with oxygen. The abundance of oxygen in the planet made it possible for a rich variety of life to evolve on earth and under the seas. The *"pikaia"* shows up, the first "fish" with *a spinal chord*, the primal ancestor of all mammals in the planet. Next, between 5 and 7 million years ago, the hominid *Homo Ergaster* appears, and becomes the common ancestor to all hominids, as shown on **Table 1** and **Figure 2**. *Homo Erectus*, a branch of Homo Ergaster, moves out of Africa and spreads throughout East Asia and Australia during the 1.4 million years, to become extinct only in modern times. A second branch, the Homo *Neanderthals* leaves Africa about 300,000 years ago and spreads throughout parts of western Europe and Asia and, again, becomes extinct in modern times, only 30,000-20,000 years ago. Still a third branch, that of *Homo Sapiens*, had prospered in Africa during the last 400,000 years, and only 50,000 years ago left Africa, initially reached Europe and the Middle East, and eventually managed to populate all the five continents. Today, all human races belong to that third branch, now called the *Homo Sapiens Sapiens* group.

Observing that the appearance and existence of modern man, Homo Sapiens Sapiens, is but a mere point in the scale of time and evolution of our planet and, realizing how numerous have been and how quickly materialize new socio-political conflicts with catastrophic potential in the last 2,000 years, it is tempting to think that our extinction awaits in the near future. In the next 1,000 years? In the next 5,000 years? In the next 20,000 years? The way we are going about it right now, the extinction date would be somewhere in that range, I would venture to say. We are not having much success in putting together better social, cultural, economic, and political models capable of anticipating and avoiding catastrophic events worldwide, some of us would say. In which case, what species would be next in reaching and claiming hegemony, supremacy in the planet. The insects? The cockroaches? The Tibetan Rabbit?

The Evolution of the Species in the Planet

The ancestors of the modern primates, our "cousins", were small and ate insects: lemurs and adapids, predominantly, as shown on **Figure 3** in the context of the last 100 million years.[5]

Boyd and **Silk (2003)** share with us part of their understanding of the circumstances in the planet during those last 100 Million years:

"In order to understand the evolutionary forces that influenced the development of the first primates we need to consider two ítems. First, what types of animals existed at the time upon which natural selection would operate? Second, what type of animal would most likely succeed in its evolution?...The plesiadapiforms varied in size, from very small and looking like a shrew, to larger animals like a marmot, and although most of these species are known today due to their teeth, they were very likely four-legged, solitary, and nocturnal...The *Eocene period* (some 55-35 million years ago) was more humid and hotter that the preceding Pelocen period, with large tropical forests covering much of the planet... In those Eocenic primates we can see for the first time characteristics of modern primates... The oldest hominoids were members of the Precunsal gender. That gender includes five species that go from the size of a macaque (10 kgs.), to the size of an obo monkey (38 kgs.). The oldest fossils found in Losidok, north of Kenya, go back 27 million years, while other fossils found in Africa date back to times as recent as 17 million years ago." [5]

Do we humans come from different places in the planet, and from different morphologic models? Again, scientific discoveries in the last 30 years tell us the following:

"A dramatic change in the morphology of the hominids took place during the glacial period. Some 100,000 years ago the globe was inhabited by a collection of hominids morphologically similar: Neanderthals in Europe, other robust hominids in east Asia, and humans a bit more modern in the Middle East... **30 years ago** a majority of paleoanthropologists would have given the same answer: the robust hominids from the end of the Superior Pleistocene were part of the same species (Homo sapiens archaic), from which gradually evolved the modern morphology that we have in the planet today.

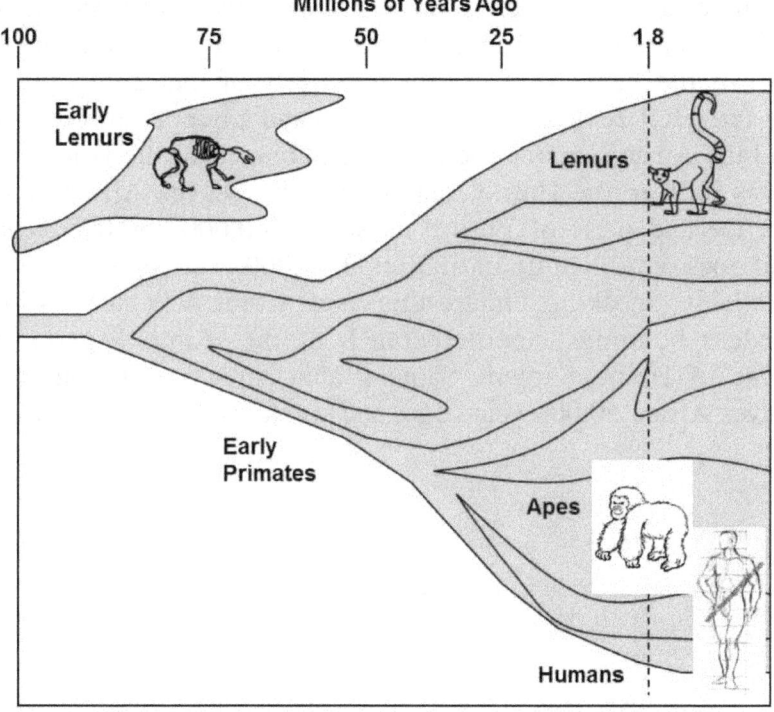

Millions of Years Ago

Figure 3. Evolution of the Species on Planet Earth. [5]

That idea was baptized with the name of **multi-regional hypothesis**... The **new evidence**, however, point to genes from an African population between 200,000 and 100,000 years ago that gave shape to the modern morphology. The individuals that carried these genes later spread all over Africa, giving birth to a large number of populations with modern morphology but genetically diverse. Then, some 50,000 years ago, *a few individuals from one of those populations* left Africa and spread all over the globe, replacing the other populations of hominids with relatively little genetic content in them... A great deal of the information comes from the genes that reside inside the *mitochondria*...genes that are *inherited via the females* only."[6]

A rich genetic variety in the Homo Sapiens Sapiens that originated from one population in Africa some 50,000 ago, as illustrated in *Figure 4.* We summarize then, saying that the hominid group called *Homo Ergaster* originated in Africa 1.8 million years ago, and spread to several regions in that continent and the Middle East. Next, a genetic variety of that group, the *Homo Erectus*, left Africa 1.6 million years and extended throughout Asia and the Australian continent. Another branch of the Homo Ergaster gives birth to the Homo Neanderthal that leaves Africa and enters Europe and parts of Euro-Asia some 400,000-350,000 years ago, and goes on to live up until 30,000-40,000 years ago, very recently, relatively speaking. Interesting and fortunately for us all, the modern hominids, another branch of the Homo Ergaster by the name of Homo Sapiens Sapiens also spreads throughout Africa, leaves Africa 50,000 years ago, and spreads all over the planet.

Do we know that we humans have been evolving during 2,000.000 years and more?

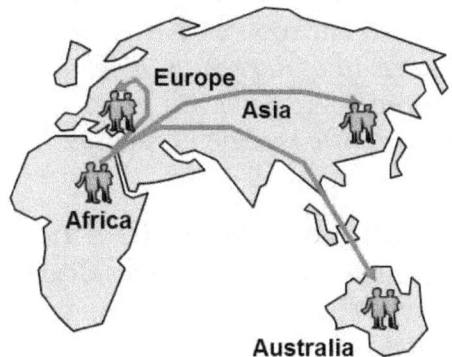

Migration and Evolution of all Human Species:

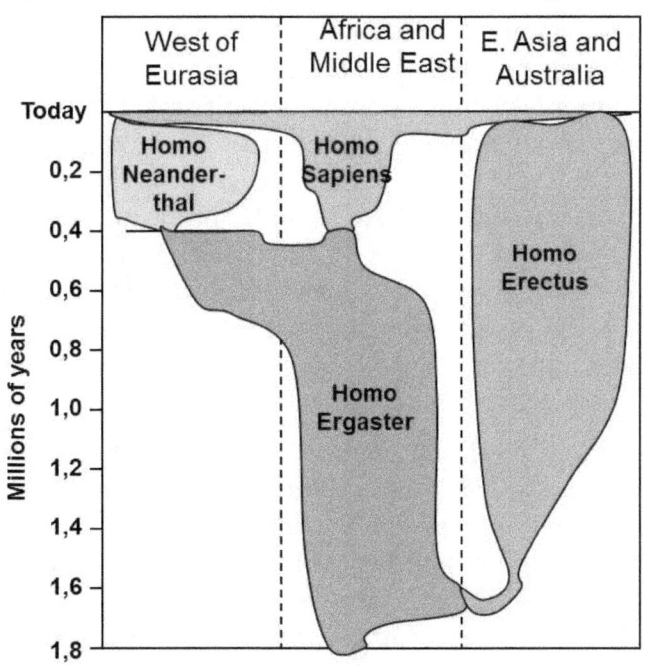

Figure 4. Migration of the Hominids out of Africa

.

One more time, all races in the planet originated with Homo Sapiens Sapiens, that group of hominids made up of a few dozens of families that left Africa (area of Ethiopia) 50,000 years ago, spreading and reaching all five continents. Greeks, Tibetans, Galicians, Germans, Catalans, Russians, Native Americans, **Basques**, Rumanians, Jews, Spaniards, Irish, Palestinians, Moais... all. We all have our roots in Africa, we all come from Africa. *Our ancestors were all of dark skin, black and curly hair. Let us review our beliefs and perspectives in order to reflect this incredible and marvelous scientific reality*, I propose.

Leakey's contributions to Paleontology [Return]

In recent times the paleontologist Richard Leakey (1944, Nairobi, Kenya-) has made significant contributions to the origins of man through his discovery of numerous fossils in East Africa:

> The son of noted anthropologists Louis S.B. Leakey and Mary Leakey, Richard was originally reluctant to follow his parents' career and instead became a safari guide. In 1967 he joined an expedition to the Omo River valley in *Ethiopia*. It was during this trip that he first noticed the site of Koobi Fora, along the shores of Lake Turkana (Lake Rudolf) in Kenya, where he led a preliminary search that uncovered several stone tools. From this site alone in the subsequent decade, Leakey and his fellow workers uncovered some 400 *hominid fossils* representing perhaps 230 individuals, making Koobi Fora the site of the richest and most varied assemblage of early human remains found to date anywhere in the world.

Leakey proposed controversial interpretations of his fossil finds. In two books written with science writer Roger Lewin, *Origins* (1977) and *People of the Lake* (1978), Leakey presented his view that, some 3 million years ago, three hominid forms coexisted: *Homo habilis*, *Australopithecus africanus*, and *Australopithecus boisei*. He argued that the two australopith forms eventually died out and that *H. habilis* evolved into *Homo erectus*, the direct ancestor of *Homo sapiens*, or modern human beings. He claimed to have found evidence at Koobi Fora to support this theory. Of particular importance is an almost completely reconstructed fossil skull found in more than

300 fragments in 1972 (coded as KNM-ER 1470). Leakey believed that the skull represented *H. habilis* and that this relatively large-brained, upright, bipedal form of *Homo* lived in eastern Africa as early as 2.5 million or even 3.5 million years ago. Further elaboration of Leakey's views was given in his work *The Making of Mankind* (1981).

From 1968 to 1989 Leakey was director of the National Museums of Kenya. In 1989 he was made director of the Wildlife Conservation and Management Department (the precursor to the Kenya Wildlife Service [KWS]). Devoted to the preservation of Kenya's wildlife and sanctuaries, he embarked on a campaign to reduce corruption within the KWS, crack down (often using force) on ivory poachers, and restore the security of Kenya's national parks. In doing so he made numerous enemies. In 1993 he survived a plane crash in which he lost both his legs below the knee. The following year he resigned his post at the KWS, citing interference by Kenyan President Daniel in the Moi's government, and became a founding member of the opposition political party Safina (Swahili for "Noah's ark"). Pressure by foreign donors led to Leakey's brief return to the KWS (1998–99) and to a short stint as secretary to the cabinet (1999–2001). Thereafter he dedicated himself to lecturing and writing on the conservation of wildlife and the environment. Another book with Roger Lewin was *The Sixth Extinction: Patterns of Life and the Future of Humankind* (1995), in which he argued that human beings have been responsible for a catastrophic reduction in the number of plant and animal species living on the Earth. Leakey later collaborated with Virginia Morell to write his second memoir, *Wildlife Wars: My Fight to Save Africa's Natural Treasures* (2001; his first memoir, *One Life*, was written in 1983). In 2004 Leakey founded Wildlife Direct, an Internet-based nonprofit conservation organization designed to disseminate information about endangered species and to connect donors to conservation efforts. He also served in

2007 as interim chair of the Kenya branch of Transparency International, a global coalition against corruption.

Leakey's wife, zoologist Meave Leakey (née Epps), conducted numerous paleoanthropological projects in the Turkana region, often in collaboration with their daughter Louise (b. 1972). In 1998 her team discovered fossil remains, more than three million years old, of a hominid that she named *Kenyanthropus platyops*. [10]

--There you have it. The origins of mankind. But, as many of you in the audience have indicated, it is not going to be easy to communicate this knowledge to our diverse population in the world. Are there any other questions so far?

--Well yes. –Said a young lady sitting in the front row at the conference room – How does the IWP propose to harmonize relations among some of the world communities with different religions and which have been engaged or are currently engaged in warfare?

--Good question! But it is also time to ask my partner **Kathy Thompson** to take over and continue with the next set of slides and questions – Xabier is now looking and smiling at Kathy sitting in the audience – Kathy, we are now in your hands!

--Thank you, Xabier, I'll do my best with the second half of our presentation on Scenario 5. Hello everyone, my name is Kathy Thompson, a member of the research team at IWP, and I will do my best to answer your questions as we go along. Yes, currently we have two major war and guerrilla activity going on, namely: (1) the ISIL (Islamic State of Iraq and the Levant) and its terrorist activities, and (2) the guerrilla activity between Israel and the Palestine people. Are we ready? Ok, let us take a look at the next PowerPoint slide.

(PowerPoint slide 23):

Caliphates and Civil Wars [Return]

The beginning of a first Islamic State. In 622 in the city of Medina, with his new converts, Mohammed created an Islamic State and drafted its own constitution, the Constitution of Medina. It was an agreement between Mohammed and all the significant tribes and families of Medina, including Muslims, Jews, Christians, and pagans to bring to an end all inter-tribal fighting, instituting a number of rights to each of those communities into a single larger community, the **Ummah**. Religious freedoms, the security of women, stable tribal relations within the Ummah, a tax system to support the community in time of conflict, and a judicial system for resolving disputes. Non-Muslims could use their own laws. Wars in the horizon. Within a few years two battles were fought against the Mecca forces, the *battle of Badr* in 624, a Muslim victory, and the *battle of Uhud* a year later, when the Meccans returned to Medina.

Mohammed dies in 632. Disagreement breaks out over who would succeed him as leader of the new community. A companion and close friend of Mohammed, **Abu Bakr**, was chosen to be the first **caliph**. Under his mandate the Muslims expanded to *Syria*, and the Khoran was compiled into a single volume by the followers of Mohammed. Other wars of Islamic expansion would soon follow. Abu Bakr dies in 634 and is succeeded by **Umar Ibn al-Khattab** as the new caliph. **Uthman ibnal-Affan, Ali ibn Abi Talib**, and **Hasan ibn Ali** would follow in that order as caliphs, and under their command the new Muslim forces expanded into **Persian** and **Byzantine** lands. Umar was assassinated by Persians in 644, Uthman was killed in a battle, and Ali was also assassinated in 1661 during the struggle known as the *first civil war*, the first *fitna*.

Next, **Mu'awiyah** came to power and began the **Umayyad Dynasty**. Soon disputes over beliefs and power emerged creating schism in the Muslim community. A large group of followers accepted the legitimacy of the three rulers prior to Ali (i.e., Abu, Umar, and Uthman) and constituted themselves into a group known as the Sunnis. A smaller group of followers disagreed, believing that only Ali and his descendants should rule, and they became known as the Shia. With Mu'awiyah's death in 680 civil war breaks out again, known as the **second Fitna**. More territorial expansion. The *Umayyad Dynasty* goes on to conquer the **Maghreb** (today's regions of Morocco, Algeria, Tunisia, Mauritania, and Libya, that is most of Northwest Africa), the **Iberian Peninsula, Narbonnese Gaul** (i.e., a Roman province located in what is now Languedoc and Provence, in southern France), and **Sindh** (i.e., one of four provinces in today's Pakistan.). Sure enough, in those regions the Muslim armies received support from local populations of Jews and Christians being persecuted as religious minorities and heavily taxed by their respective local powers. During the Al-Andalus era in Spain, the Maghreb's inhabitants, Maghrebis, were known as *"Moors"*.

More conquests. Descendants of Abbas ibn Abd al-Muttalib, Mohammed's uncle, were able to gather non-Arab converts called *mawali,* poor Arabs, and Shi'a against the Umayyads and overthrew them, thus creating the *Abbasid Dynasty* in 750.[2] Sure enough, during this dynasty the Delhi Sultanate took over the Indian subcontinent, and Muslims went to China to trade dominating much of the import and export industry over the domains of the Chinese *Song Dynasty*.[3]

What about doctrines and philosophies? To begin with, many *hadith* collections were compiled during the Abbasid period. Al-Tabari and Ibn Kathir completed the writing of popular commentaries of the Khoran. Philosophers Al-Farabi and Avicenna tried to incorporate some Greek principles into Islamic theology, while Al-Ghazali opposed that movement and ultimately succeeded. Ascetics such as Hasan al-Basri created the Sufism movement, a movement that later in the 13th century would evolve into the *Sufi Order* of spiritual teachers and students.[4]

Islam's Golden Age [Return]

By the 9th century Islam is able to create outstanding institutions and achieve notable goals. The *University of Al Karaouine* (Sunni, in today's Fes, Morocco) was founded in 859 and went on to become the world's oldest degree-granting university. Ibn *Al-Haytham* (965-1040) has been regarded by many as the father of the *modern scientific method*. *Al-Jahiz* (776-869) reportedly proposed a theory of *natural selection* after observing physical and mental traits of the *Zanj* black tribes.[12] It was in 1258 when the *Mongol empire* put an end to that Abbasid Dynasty.

Modern Times (1258-Present) [Return]

Islam continued to expand into Europe, Sub-Saharan Africa, Central Asia, and the Malay islands:

> The Arab conquests in *Spain* followed from their rapid sweep through North Africa in the seventh century. Conquests of Spain were complete, save for a Christian remnant in the northeast, by 732 CE. Frankish armies defeated the Muslims in France, blocking further gains; and a brief hold over Sicily and other Italian islands was pushed back by Christian invaders. But the Muslim period in Spain and *Portugal* had vital consequences. Muslim rulers developed an elaborate political and cultural framework while largely tolerating Christian subjects. A number of Spaniards converted under the influence of conquest and Muslim success. Muslim artistic styles long influenced Spanish architecture and decoration, even after Islam itself had been pushed out. Music, including the guitar, an Arab instrument, merged traditions as well—and from Spain the new styles would later spread to *the Americas*. Centers of learning, like Toledo, drew scholars from all over Europe, eager to take advantage of Muslim and Jewish *science and philosophy*; the result helped spur change and development in European intellectual life. Amid all this fruitful interaction, Christian warriors from northern Spain mounted a steady counterattack, gradually winning back territory from the tenth century onward. The strength of Christianity and, ironic ally, limited trade opportunities in backward Europe prevented the spread of Muslim influence, and the retreat was inexorable, particularly as Arab political consolidation in the Middle East and Africa broke down, leaving the rulers in Spain isolated. In1492 CE the last remaining pocket, in Granada, was expelled by the forces of the now -united Spanish monarchy of Ferdinand and Isabella. [13]

Under the Ottoman empire Islam was able to spread to Southeast Europe, Crimea, and the Caucasus. However, by the 1800's the Muslim world was in serious cultural decline, without a major observatory by the 20th Century. The British Empire had also managed to end the Mughal dynasty in India by the 19th Century. *Liberal Islam* was able to reconcile religious traditions with modern norms and human rights in a number of governments and countries. *Women issues* receive significant weight in modern times as well. [16] An *Islamic revival* has taken place:

> Jamal-al-Din al-Afghani, along with his acolyte Muhammad Abduh, have been credited as forerunners of the Islamic revival. Abul A'la Maududi helped influence *modern political Islam*. Islamist groups such as the Muslim Brotherhood advocate Islam as a comprehensive political solution, often in spite of being banned. In Iran, revolution replaced a secular regime with an *Islamic state*. In Turkey, the Islamist AK Party has democratically been in power for about a decade, while Islamist parties did well in elections following the *Arab Spring*. The Organization of Islamic Cooperation (OIC), consisting of Muslim countries, was established in 1969 after the burning of the Al-Aqsa Mosque in Jerusalem. Piety appears to be deepening worldwide. In many places, the prevalence of the *Islamic veil* is growing increasingly common and the percentage of Muslims favoring *Sharia laws* has increased. With religious guidance increasingly available electronically, Muslims are able to access views that are strict enough for them rather than rely on state clerics who are often seen as stooges. Some organizations began using the media to promote Islam such as the 24-hour TV channel, Peace TV. Perhaps as a result of these efforts, most experts agree that Islam is growing faster than any other faith in East and West Africa. [9][14]

Sunni and Shia [Return]

As we may already know, there are the main Islamic groups: (1) the Sunni, and (2) the Shia. *Sunnis* believe that the first four caliphs were the rightful successors to Muhammad, since God did not specify any particular leaders to succeed him. Sunnis believe that anyone who is righteous and just could be a caliph but they have to act according to the Qur'an and the Hadith, in that order of priority. The Sunni make up 75%-90% of all Muslims.

Shia Islam has several branches, the largest of which is the *Twelvers*, followed by Zaidis and Ismailis:

> The Shia constitute 10%–20% of Islam and are its second-largest branch. While the Sunnis believe that a Caliph should be elected by the community, Shia's believe that Muhammad appointed his son-in-law, Ali ibn Abi Talib, as his successor and only certain descendants of Ali could be Imams. As a result, they believe that Ali ibn Abi Talib was the first *Imam* (leader), rejecting the legitimacy of the previous Muslim caliphs Abu Bakr, Uthman ibn al-Affan and Umar ibn al-Khattab. Shia Islam has several branches, the largest of which is the Twelvers, followed by Zaidis and Ismailis. Different branches accept different *descendants of Ali* as Imams. After the death of Imam Jafar al-Sadiq considered the sixth Imam by the Twelvers, and the Ismaili's, the Ismailis started to consider his son Isma'il ibn Jafar as the Imam and the Twelver Shia's (Ithna Asheri) started to consider his other son Musa al-Kazim as their seventh Imam. While the Zaydis consider Zayd ibn Ali, the uncle of Imam Jafar al-Sadiq, as their fifth Imam. Other smaller groups include the Bohra and Druze, as well as the Alawites and Alevi. Some Shia branches label other Shia branches that do not agree with their doctrine as Ghulat. [15]

Arab-Israeli Conflict

The *International law* bearing on issues of *Arab–Israeli conflict*, have become a major arena of regional and international tension since the birth of Israel in 1948, resulting in several disputes between a number of Arab countries and Israel. For example, as a result of the *Six-Day War in 1967*, Israel has come to occupy land invaded and occupied in 1948 by neighboring Egypt, Syria and Jordan. Following the peace treaties between Israel and Egypt and Israel and Jordan, the conflict today largely revolves around Palestinian statehood.

Main points of dispute (also known as the "core issues" or "final status issues") are the following:

- the legality of the Israeli settlements in the Palestinian territories, and their annexation of East Jerusalem and the Israeli West Bank barrier;
- how legal borders should be decided between Israel and a Palestinian state;
- the legal status of the Palestinian refugees from the 1948 Arab–Israeli war and subsequently.

The *United Nations (UN) General Assembly* has voted on a resolution bearing on issues of international law as applied to the conflict *every year* since 1974. And yet, the conflict goes on without a solution.

Coordination of Efforts through IWP

--So what is IWP proposing in order to resolve these two major conflicts?

--Agree. There is a great and immediate need to address these two main issues and, to that effect, IWP is proposing coordinating its efforts with the WCR and the United Nations (UN), as the following PowerPoint slide proposes:

(PowerPoint slide 25):

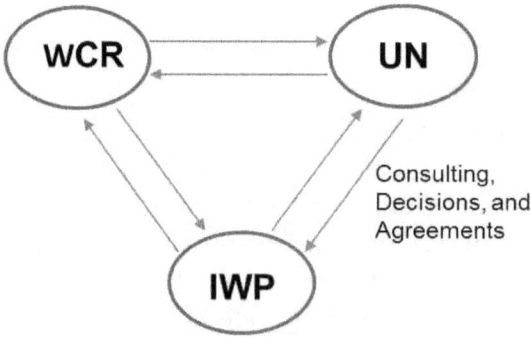

Figure 5. Coordination among the World Council of all Religions (WCR), the International Organization for World Peace (IWP), and the United Nations (UN)

Major considerations in this coordinating effort falls into 3 areas: (1) IWP and UN, such that IWP provides information to UN on communities and countries which may be falling below economic, social, and political thresholds, as to then try to prevent an outbreak of war, (2) WCR and IWP, so that WCR considers areas of potential war conflict and would try to keep religion and politics from interfering with each other, and (3) UN and WCR, so that after a war or guerrilla outbreak between two countries, the UN would be able to continue sending security forces to the affected area with the support from WCR, thus preventing "religious wars."

A hand is raised on the second row of the conference room.

--Maybe this is just me, but I see many complications in the process of creating and effective World Council of all Religions (WCR) and a coordinating effort between it and the other two international entities, IWP and the UN. Too many agreements and changes to historical ways of thinking among human beings would have to take place. What is the *probability of success* that you would assign to this *Scenario 5* actually happening?

--Of course, we are going to try to come up with an answer to that question, but for that purpose we are going to ask *Dr. Eugene Finley* to participate, as he is more familiar with all the scenarios. Dr. Finley?

--Thank you, Kathy! Glad to be here today. For one, I am impressed with the list of very good questions which all of you in the audience have asked Xavier and Kathy. Great job! Well, to get started with the answer I should say that "*this and the other 4 scenarios are not mutually exclusive*", which is to say that two or more scenarios can occur simultaneously, and that would complicate the situation in our planet considerably. Regarding the probability of Scenario happening and its success, I would have to say that it is low, very low, in the order of 10%-15% only. In fact, I would venture to say that any scenario in the category of "*negotiate-and-live*" has a lower probability of happening and succeeding than that of any scenario in the category of "*fight-and-die*." Why is that so? Well, this is something which we plan to address in the following 3 scenarios, for sure. Meantime, I would like to thank Kathy, Xabier, and all of you in the audience for your participation and interest. See you next week for the next Scenario. Ohh, sorry, please remember to complete the one-page survey on your tables with your comments and suggestions to our IWP organization.

Thank you all!

***

Chapter 10:

Scenario 6: Promote Economic and Political Balance across the Global Community

*"We cannot change **the political system**, we cannot change **the economic system**, we cannot change the social system, until the people control the land, and then we take it out of the hands of that sick minority that chooses to pervert the meaning and the intention of humanity."*

*"**Growing economies are critical**; we will never be able to end poverty unless economies are growing. We also need to find ways of growing economies so that the growth creates good jobs, especially for young people, **especially for women**, especially for the poorest who have been excluded from the economic system."*
*-- **Jim Yong Kim***
Read more at:

https://www.brainyquote.com/quotes/keyword
s/economic_system.html

--Welcome to our presentation of *Scenario 6* at the University of Arizona, here in Tucson, Arizona, USA. As you all know this scenario series is sponsored by the *International World Organization for Peace (IWP)* in an effort to prepare our community of nations in the world for the difficult times which are coming over the next 250 years on the topics of overpopulation, climatic change, the danger of nuclear warfare, and the threat of religious warfare, to mention a few dimensions. Yes, my name is *Kathy Thompson*, I will be doing the first half of this presentation, to be followed by *Xabier Elurmendi*, also a member of the IWP research team. Any questions before I begin to share with you this Scenario 6 with title:

"Promote Economic and Political Balance across the Global Community"

in this conference room?

Three hands are raised in the air in the audience.

--Yes, the person in the last row. What is your question, please?

--Thank you. I was hoping you would be able to tell us what other organizations are collaborating with IWP in this seminar series. Again, thanks.

--Well, for one, we have some members of the *United Nations (UN)* sharing with us their views and goals on Economic Balance, the *International Monetary Fund (IMF)* with its perspective on "globalization", and our own *International World Organization for Peace (IWP)*, of course.

--Thank you!

--Yes, the hand on the second row, please.

--What are the major elements of this Scenario 5 which make it an *"extreme scenario"*? If you could briefly give us such a list, please.

--Of course, I will be glad to do so, and for that purpose I will begin the presentation with such a list of major elements or "main factors." But first, I need to respond to the third hand up in the air on the first row. Yes, your question, please?

--Thank you. Yes, I wanted to ask you about the probability of this scenario happening, and over which period of time. Are we talking about something about to happen in the next 25-50 years, or in the next 500 years?

--Also a very good question, and we will address it in the very first few PowerPoint slides, of course. Please bear with me for a few minutes, and we will then get back to your question, OK? Thank you.

Main Factors in Scenario 5

--OK, to get started then, let us take a look at the first PowerPoint slide on the *main factors* of this *Scenario 5*:

- Large number of Religions in the World.
- Atheist organizations in the community
- Lack of Economic balance among nations.
- Lack of Political balance among nations
- Vulnerable populations in poverty
- Globalization process
- Major countries seeking "rupture" from traditional partnerships with other countries.

--Several items to bring up at this time in relation to these main factors. For one, the IWP organization would like us to recognize that we humans, as a species, *as primates that we are*, have already built into our genes over millions of years, the desire to survive and to acquire materials goods, as much as we can. Meaning that even though our world community may be in a period of "relative peace", some of our communities and nations want more land, more commerce with other nations, more political, and more military power. It is also true that a very large percentage of our global population exists with very low economic means, compared to a few rich nations, that is economically and politically very rich nations. This is a main reason why some opportunistic political leaders are successful at gathering a large number of followers, promising better living conditions and good jobs, to try to disrupt current political systems. As long as we have a great disparity between a few rich people and a very large percentage of poor people in a country, or group of countries, this desire to become involved in political and military movements will materialize in the form of street demonstrations and eventual guerrilla or warfare activity. I just have to mention the case of **Hitler** in Germany, **Franco** in Spain, **Mussolini** in Italy, during World War II, President **Maduro** in Venezuela, and the dictator **Kim Jong** in North Korea today to make my point.

Also, and most important in this Scenario 5, is the tendency of our human species, **Homo Sapiens Sapiens**, to continue looking, searching for new horizons, new lands, new discoveries, roaming around our planet as we have been doing thousands of years, possibly millions of years. Therefore, this is a scenario where we contemplate the "**crossroad**", the confluence, of our desires and capabilities as a species, on one hand, and the limited space of our planet, on the other hand.

Goals of the United Nations (UN)

Is everyone ready to know about the UN goals in areas related to this Scenario 5? OK, then let's go to PowerPoint slides 1 and 2 showing a condensed view of this scenario, and then proceed with a summary of those UN goals.

GOAL 1: End Poverty everywhere[1]

- *860 Million in the planet still live under extreme poverty conditions.*

- *About 1 in 5 persons in developing regions lives on less than $1.25 per day.*

- *The overwhelming majority of people living on less than $1.25/day belong to the regions of Southern Asia and sub-Saharan Africa.*

- *High poverty rates are often found in small, fragile and conflict-affected countries.*

- *One in four children under age five in the world has inadequate height for his or her age.*

- *Every day in 2014, 42,000 people had to abandon their homes to seek protection due to conflict.*

(PowerPoint slide No. 1):

Main Factors:
- Large number of Religions in the World.
- Atheist organizations in the community
- Lack of Economic balance among nations.
- Lack of Political balance among nations
- Vulnerable populations in poverty
- Globalization process
- Major countries seeking "rupture" from traditional parternerships with other countries.

Time Frame: The following 500 years.

United Nations (UN) Economic and Social Goals:
- Goal 1: End Poverty.
- Goal 2: End Hunger, sustainable agriculture.
- Goal 3: Promote Healthy lives.
- Goal 4: Quality Education for all peoples.
- Goal 5: Achieve Gender Equality and empowerment for Women
- Goal 6: Access to Water and sanitation for all.
- Goal 7: Access to Energy for all families.
- Goal 8: Promote Economic growth and employment
- Goal 9: Develop infrastructures, industrialization and foster innovation
- Goal 10: Reduce Inequality among Nations.
- Goal 11: Develop safe cities.
- Goal 12: Ensure sustainable Consumption.
- Goal 13: Prepare for Climate Change (CC).
- Goal 14: Protect and develop Sea Resources.
- Goal 15: Manage Forests and Biodiversity.
- Goal 16: Promote just and inclusive societies.
- Goal 17: Revitalize Global Partnership.

Figure 1. Scenario 6, Development of Economic and Political Balance accross the Global Community   (Part 1 of 2)

(PowerPoint slide No. 2):

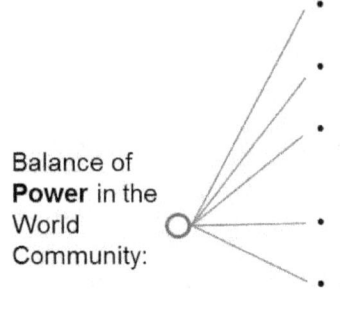

Balance of **Power** in the World Community:

- Goals, Means, and Dynamics of Balance of Power
- Vigilance in the distribution of political and military capabilities among Nations.
- Association among groups of Nations to create "poles" of political and military power.
- In times of conflict, powers to appeal to thinking, compromises, avoid wars.
- As a last resort, pursue moderate war aims.

Countries with **lower labor costs** would benefit.
Employment in countries with lower labor costs would improve considerably.
International trade increases.
It would contribute to income equality among countries.
On the other hand, **large companies and interests** would dictate the economy of countries.
Citizen organizations in some countries (e.g., USA, France, United Kingdom, others) **would want to impede** globalization.

The **Good** and the **Bad** side of Globalization.

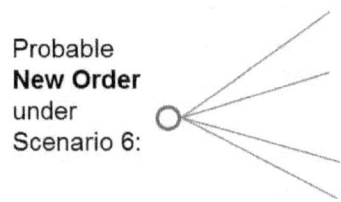

Probable **New Order** under Scenario 6:

- Rise of the **average level of employment** among nations.
- Ideally, improved living conditions would **detract major "polar" countries from provoking wars**.
- Reduce **poverty** worlwide.
- **Promote education** and development of skills among developing countries.

Figure 2. Scenario 6, Development of Economic and Political Balance accross the Global Community   (Part 2 of 2)

Poverty does not mean access to food supplies only and, instead, it means lack to access to social and educational opportunities in a community, as **Figure 3** implies.

Figure 3. Poverty in our communities includes hunger, malnutrition, social descrimination, and limited access to education. [1]

GOAL 2: End hunger[1]

- *Globally, one in nine people in the world today (795 million) are undernourished.*

- *The vast majority of the world's hungry people live in developing countries, where 12.9 per cent of the population is undernourished.*

- *Asia is the continent with the most hungry people – two thirds of the total. The percentage in southern Asia has fallen in recent years but in western Asia it has increased slightly.*

- *Southern Asia faces the greatest hunger burden, with about 281 million undernourished people. In sub-Saharan Africa, projections for the 2014-2016 period indicate a rate of undernourishment of almost 23 per cent.*

- *Poor nutrition causes nearly half (45 per cent) of deaths in children under five – 3.1 million children each year.*

GOAL 3: Provide for healthy lives and well-being[1]

- *17,000 fewer children die each day than in 1990, **but** more than six million children still die before their fifth birthday each year*

- *Since 2000, measles vaccines have averted nearly 15.6 million deaths*

- *Despite determined global progress, an increasing proportion of child deaths are in sub-Saharan Africa and Southern Asia. Four out of every five deaths of children under age five occur in these regions.*

- *Children born into poverty are almost twice as likely to die before the age of five as those from wealthier families.*

- *Children of educated mothers—even mothers with only primary schooling—are more likely to survive than children of mothers with no education.*

Maternal health:

- *Maternal mortality has fallen by almost 50 per cent since 1990*

- *In Eastern Asia, Northern Africa and Southern Asia, maternal mortality has declined by around two-thirds*

- *But maternal mortality ratio – the proportion of mothers that do not survive childbirth compared to those who do – in developing regions is still 14 times higher than in the developed regions*

- *More women are receiving antenatal care. In developing regions, antenatal care increased from 65 per cent in 1990 to 83 per cent in 2012*

- *Only half of women in developing regions receive the recommended amount of health care they need*

- *Fewer teens are having children in most developing regions, but progress has slowed. The large increase in contraceptive use in the 1990s was not matched in the 2000s*

- *The need for family planning is slowly being met for more women, but demand is increasing at a rapid pace*

HIV/AIDS, malaria and other diseases:

- *At the end of 2014, there were 13.6 million people accessing antiretroviral therapy*

- *New HIV infections in 2013 were estimated at 2.1 million, which was 38 per cent lower than in 2001*

- *At the end of 2013, there were an estimated 35 million people living with HIV*

- *At the end of 2013, 240 000 children were newly infected with HIV*

- *New HIV infections among children have declined by 58 per cent since 2001*

- *Globally, adolescent girls and young women face gender-based inequalities, exclusion, discrimination and violence, which put them at increased risk of acquiring HIV*

- *HIV is the leading cause of death for women of reproductive age worldwide*

- *TB-related deaths in people living with HIV have fallen by 36% since 2004*

- *There were 250 000 new HIV infections among adolescents in 2013, two thirds of which were among adolescent girls*

- *AIDS is now the leading cause of death among adolescents (aged 10–19) in Africa and the second most common cause of death among adolescents globally*

- *In many settings, adolescent girls' right to privacy and bodily autonomy is not respected, as many report that their first sexual experience was forced*

- *As of 2013, 2.1 million adolescents were living with HIV*

- *Over 6.2 million malaria deaths have been averted between 2000 and 2015, primarily of children under five years of age in sub-Saharan Africa. The global malaria incidence rate has fallen by an estimated 37 per cent and the mortality rates by 58 per cent*

- *Between 2000 and 2013, tuberculosis prevention, diagnosis and treatment interventions saved an estimated 37 million lives. The tuberculosis mortality rate fell by 45 per cent and the prevalence rate by 41 per cent between 1990 and 2013.*

GOAL 4: Promote quality education[1]

- *Enrolment in primary education in developing countries has reached 91 per cent but 57 million children remain out of school.*

- *More than half of children that have not enrolled in school live in sub-Saharan Africa.*

- *An estimated 50 per cent of out-of-school children of primary school age live in conflict-affected areas.*

- *103 million youth worldwide lack basic literacy skills, and more than 60 per cent of them are women.*

GOAL 5: Achieve gender equality and empowerment for Women[1]

- *About two thirds of countries in the developing regions have achieved gender parity in primary education.*

- *In Southern Asia, only 74 girls were enrolled in primary school for every 100 boys in 1990. By 2012, the enrolment ratios were the same for girls as for boys.*

- *In sub-Saharan Africa, Oceania and Western Asia, girls still face barriers to entering both primary and secondary school.*

- *Women in Northern Africa hold less than one in five paid jobs in the non-agricultural sector. The proportion of women in paid employment outside the agriculture sector has increased from 35 per cent in 1990 to 41 per cent in 2015.*

- *In 46 countries, women now hold more than 30 per cent of seats in national parliament in at least one chamber.*

Empowerment for women means access to higher education, participation in product development, and visible representation in the research community, as shown on *Figure 4*.

Figure 4. Education for Women and their empowerment in our societies is vital to our global society. [1]

GOAL 6: Ensure access to water and sanitation for all[1]

- *2.6 billion people have gained access to improved drinking water sources since 1990, but 663 million people are still without.*

- *At least 1.8 billion people globally use a source of drinking water that is fecally contaminated.*

- *Between 1990 and 2015, the proportion of the global population using an improved drinking water source has increased from 76 per cent to 91 per cent.*

- *But water scarcity affects more than 40 per cent of the global population and is projected to rise. Over 1.7 billion people are currently living in river basins where water use exceeds recharge.*

- *2.4 billion people lack access to basic sanitation services, such as toilets or latrines.*

- *More than 80 per cent of wastewater resulting from human activities is discharged into rivers or sea without any pollution removal.*

- *Each day, nearly 1,000 children die due to preventable water and sanitation-related diarrheal diseases.*

- *Hydropower is the most important and widely-used renewable source of energy and as of 2011, represented 16 per cent of total electricity production worldwide.*

- *Approximately 70 per cent of all water abstracted from rivers, lakes and aquifers is used for irrigation.*

- *Floods and other water-related disasters account for 70 per cent of all deaths related to natural disasters.*

GOAL 7: Access to affordable energy[1]

- *One in five people still lacks access to modern electricity.*

- *3 billion people rely on wood, coal, charcoal or animal waste for cooking and heating.*

- *Energy is the dominant contributor to climate change, accounting for around 60 per cent of total global greenhouse gas emissions.*

- *Reducing the carbon intensity of energy is a key objective in long-term climate goals.*

Oil resources may last another 25-50 years, at most, making necessary the creation of alternative energy sources, including solar and wind sources, as Figure 5 implies.

Figure 5. Wind energy and other alternative energy sources are essential and necessary in our Cities and towns are to advance in the future. [1]

GOAL 8: *Promote sustainable economic growth and employment*[1]

- *Global unemployment increased from 170 million in 2007 to nearly 202 million in 2012, of which about 75 million are young women and men.*

- *Nearly 2.2 billion people live below the US$2 poverty line and that poverty eradication is only possible through stable and well-paid jobs.*

- *470 million jobs are needed globally for new entrants to the labour market between 2016 and 2030.*

GOAL 9: Promote Industrialization and Innovation[1]

- *Basic infrastructure like roads, information and communication technologies, sanitation, electrical power and water remains scarce in many developing countries.*

- *About 2.6 billion people in the developing world are facing difficulties in accessing electricity full time*

- *2.5 billion people worldwide lack access to basic sanitation and almost 800 million people lack access to water, many hundreds of millions of them in Sub Saharan Africa and South Asia*

- *1-1.5 billion people do not have access to reliable phone services*

- *Quality infrastructure is positively related to the achievement of social, economic and political goals*

- *Inadequate infrastructure leads to a lack of access to markets, jobs, information and training, creating a major barrier to doing business*

- *Undeveloped infrastructures limits access to health care and education*

- *For many African countries, particularly the lower-income countries, the existent constraints regarding infrastructure affect firm productivity by around 40 per cent*

- *Manufacturing is an important employer, accounting for around 470 million jobs worldwide in 2009 – or around 16 per cent of the world's workforce of 2.9 billion. In 2013, it is estimated that there were more than half a billion jobs in manufacturing*

- *Industrialization's job multiplication effect has a positive impact on society. Every one job in manufacturing creates 2.2 jobs in other sectors*

- *Small and medium-sized enterprises that engage in industrial processing and manufacturing are the most critical for the early stages of industrialization and are typically the largest job creators. They make up over 90 per cent of business worldwide and account for between 50-60 per cent of employment*

- *In countries where data are available, the number of people employed in renewable energy sectors is presently around 2.3 million. Given the present gaps in information, this is no doubt a very conservative figure. Because of strong rising interest in energy alternatives, the possible total employment for renewables by 2030 is 20 million jobs*

- *Least developed countries have immense potential for industrialization in food and beverages (agro-industry), and textiles and garments, with good prospects for sustained employment generation and higher productivity*

- *Middle-income countries can benefit from entering the basic and fabricated metals industries, which offer a range of products facing rapidly growing international demand.*

- *In developing countries, barely 30 per cent of agricultural production undergoes industrial processing. In high-income countries, 98 per cent is processed. This suggests that there are great opportunities for developing countries in agribusiness.*

As ***Figure 6*** wants to communicate, the promotion of transportation and other infrastructures in our cities are essential and necessary in order to cope with a growing population,

Figure 6. Investment in alternative transportation modes, communication technology, and research programs are needed in order to achieve sustainable societies. [1]

GOAL 10: Reduce inequality within and among countries[1]

- *On average—and taking into account population size—income inequality increased by 11 per cent in developing countries between 1990 and 2010.*

- *A significant majority of households in developing countries—more than 75 per cent of the population—are living today in societies where income is more unequally distributed than it was in the 1990s.*

- *Evidence shows that, beyond a certain threshold, inequality harms growth and poverty reduction, the quality of relations in the public and political spheres and individuals' sense of fulfilment and self-worth.*

- *There is nothing inevitable about growing income inequality; several countries have managed to contain or reduce income inequality while achieving strong growth performance.*

- *Income inequality cannot be effectively tackled unless the underlying inequality of opportunities is addressed.*

- *In a global survey conducted by UN Development Programme, policy makers from around the world acknowledged that inequality in their countries is generally high and potentially a threat to long-term social and economic development.*

- *Evidence from developing countries shows that children in the poorest 20 per cent of the populations are still up to three times more likely to die before their fifth birthday than children in the richest quintiles.*

- *Social protection has been significantly extended globally, yet persons with disabilities are up to five times more likely than average to incur catastrophic health expenditures.*

- *Despite overall declines in maternal mortality in the majority of developing countries, women in rural areas are still up to three times more likely to die while giving birth than women living in urban centers.*

GOAL 11: Make cities inclusive, safe, and sustainable[1]

- *Half of humanity – 3.5 billion people – lives in cities today.*

- *By 2030, almost 60 per cent of the world's population will live in urban areas.*

- *95 per cent of urban expansion in the next decades will take place in developing world.*

- *828 million people live in slums today and the number keeps rising.*

- *The world's cities occupy just 3 per cent of the Earth's land, but account for 60-80 per cent of energy consumption and 75 per cent of carbon emissions.*

- *Rapid urbanization is exerting pressure on fresh water supplies, sewage, the living environment, and public health.*

- *But the high density of cities can bring efficiency gains and technological innovation while reducing resource and energy consumption.*

GOAL 12: *Promote sustainable consumption and production conditions*[1]

- *Each year, an estimated one third of all food produced – equivalent to 1.3 billion tons worth around $1 trillion – ends up rotting in the bins of consumers and retailers, or spoiling due to poor transportation and harvesting practices*

- *If people worldwide switched to energy efficient lightbulbs the world would save US$120 billion annually*

- *Should the global population reach 9.6 billion by 2050, the equivalent of almost three planets could be required to provide the natural resources needed to sustain current lifestyles*

Water

- *Less than 3 per cent of the world's water is fresh (drinkable), of which 2.5 per cent is frozen in the Antarctica, Arctic and glaciers. Humanity must therefore rely on 0.5 per cent for all of man's ecosystems and fresh water needs.*

- *Man is polluting water faster than nature can recycle and purify water in rivers and lakes.*

- *More than 1 billion people still do not have access to fresh water.*

- *Excessive use of water contributes to the global water stress.*

- *Water is free from nature but the infrastructure needed to deliver it is expensive.*

Energy

- *Despite technological advances that have promoted energy efficiency gains, energy use in OECD countries will continue to grow another 35 per cent by 2020. Commercial and residential energy use is the second most rapidly growing area of global energy use after transport.*

- *In 2002 the motor vehicle stock in OECD countries was 550 million vehicles (75 per cent of which were personal cars). A 32 per cent increase in vehicle ownership is expected by 2020. At the same time, motor vehicle kilometers are projected to increase by 40 per cent and global air travel is projected to triple in the same period.*

- *Households consume 29 per cent of global energy and consequently contribute to 21 per cent of resultant CO2 emissions.*

- *One-fifth of the world's final energy consumption in 2013 was from renewables.*

Food

- *While substantial environmental impacts from food occur in the production phase (agriculture, food processing), households influence these impacts through their dietary choices and habits. This consequently affects the environment through food-related energy consumption and waste generation.*

- *1.3 billion tons of food is wasted every year while almost 1 billion people go undernourished and another 1 billion hungry.*

- *Overconsumption of food is detrimental to our health and the environment.*

- *2 billion people globally are overweight or obese.*

- *Land degradation, declining soil fertility, unsustainable water use, overfishing and marine environment degradation are all lessening the ability of the natural resource base to supply food.*

- *The food sector accounts for around 30 per cent of the world's total energy consumption and accounts for around 22 per cent of total Greenhouse Gas emissions.*

GOAL 13: Prepare for Climate Change (CC) and its negative impacts[1]

- *From 1880 to 2012, average global temperature increased by 0.85°C. To put this into perspective, for each 1 degree of temperature increase, grain yields decline by about 5 per cent. Maize, wheat and other major crops have experienced significant yield reductions at the global level of 40 megatons per year between 1981 and 2002 due to a warmer climate.*

- *Oceans have warmed, the amounts of snow and ice have diminished and sea level has risen. From 1901 to 2010, the global average sea level rose by 19 cm as oceans expanded due to warming and ice melted. The Arctic's sea ice extent has shrunk in every successive decade since 1979, with 1.07 million km² of ice loss every decade.*

- *Given current concentrations and on-going emissions of greenhouse gases, it is likely that by the end of this century, the increase in global temperature will exceed 1.5°C compared to 1850 to 1900 for all but one scenario. The world's oceans will warm and ice melt will continue. Average sea level rise is predicted as 24 – 30cm by 2065 and 40-63cm by 2100. Most aspects of climate change will persist for many centuries even if emissions are stopped*

- *Global emissions of carbon dioxide (CO2) have increased by almost 50 per cent since 1990.*

- *Emissions grew more quickly between 2000 and 2010 than in each of the three previous decades.*

- *It is still possible, using a wide array of technological measures and changes in behavior, to limit the increase in global mean temperature to two degrees Celsius above pre-industrial levels.*

- *Major institutional and technological change will give a better than even chance that global warming will not exceed this threshold.*

GOAL 15: Manage forests and biodiversity[1]

Forests

- *Around 1.6 billion people depend on forests for their livelihood. This includes some 70 million indigenous people.*

- *Forests are home to more than 80 per cent of all terrestrial species of animals, plants and insects.*

Desertification

- *2.6 billion people depend directly on agriculture, but 52 per cent of the land used for agriculture is moderately or severely affected by soil degradation.*

- *As of 2008, land degradation affected 1.5 billion people globally.*

- *Arable land loss is estimated at 30 to 35 times the historical rate.*

- *Due to drought and desertification each year 12 million hectares are lost (23 hectares per minute), where 20 million tons of grain could have been grown.*

- *74 per cent of the poor are directly affected by land degradation globally.*

Biodiversity

- *Of the 8,300 animal breeds known, 8 per cent are extinct and 22 per cent are at risk of extinction.*

- *Of the over 80,000 tree species, less than 1 per cent have been studied for potential use.*

- *Fish provide 20 per cent of animal protein to about 3 billion people. Only ten species provide about 30 per cent of marine capture fisheries and ten species provide about 50 per cent of aquaculture production.*

- *Over 80 per cent of the human diet is provided by plants. Only three cereal crops – rice, maize and wheat – provide 60 per cent of energy intake.*

- *As many as 80 per cent of people living in rural areas in dev eloping countries rely on traditional plant based medicines for basic healthcare.*

- *Micro-organisms and invertebrates are key to ecosystem services, but their contributions are still poorly known and rarely acknowledged.*

GOAL 16: Promote just, peaceful, and inclusive societies[1]

- *Among the institutions most affected by corruption are the judiciary and police.*

- *Corruption, bribery, theft and tax evasion cost some US $1.26 trillion for developing countries per year; this amount of money could be used to lift those who are living on less than $1.25 a day above $1.25 for at least six years.*

- *The rate of children leaving primary school in conflict affected countries reached 50 per cent in 2011, which accounts to 28.5 million children, showing the impact of unstable societies on one of the major goals of the post 2015 agenda: education.*

- *The rule of law and development have a significant interrelation and are mutually reinforcing, making it essential for sustainable development at the national and international level.*

GOAL 17: Promote global partnership for sustainable development[1]

- *Official development assistance stood at $135.2 billion in 2014, the highest level ever recorded.*

- *79 per cent of imports from developing countries enter developed countries duty-free.*

- *The debt burden on developing countries remains stable at about 3 per cent of export revenue.*

- *The number of Internet users in Africa almost doubled in the past four years.*

- *30 per cent of the world's youth are digital natives, active online for at least five years.*

- *But more four billion people do not use the Internet, and 90 per cent of them are from the developing world.*

--Great! We now have seen all the 17 Goals the UN has in mind as it tries to contribute to solve those many social, food, education, and industrial problems in our global community. We have seen the statistics associated with each goal, and now you can pursue each goal in greater detail on you own, of course. Any questions?

A hand is raised in the fourth row.

--Yes, I have a question. Well, it is more of a comment, actually. I have seen the statistics shown on the slides, and I also hear you say now the UN is trying to address all these problems, but these problems are too big. I cannot see how the UN or any other collection of international organizations could find solutions to such major economic disparities in the world, really.

--Yes, I understand, I hear you. We are not saying that the intent is to bring to a complete and total end those hunger and unemployment problems in the world. Instead, the UN and other international organizations are working to reduce those economic inequalities in an effort to reduce the number of deaths and wars in the world today. OK, if there are no other questions, we are now going to move to the topic of "*globalization*", and for that the next PowerPoint slide will give us some basic details:

(PowerPoint slide No. 16:)

"Economic "globalization" is a historical process, the result of human innovation and technological progress. It refers to the increasing integration of economies around the world, particularly through the movement of goods, services, and capital across borders. The term sometimes also refers to the movement of people (labor) and knowledge (technology) across international borders. There are also broader cultural, political, and environmental dimensions of globalization.

The term "globalization" began to be used more commonly in the 1980s, reflecting technological advances that made it easier and quicker to complete international transactions—both trade and financial flows. It refers to an extension beyond national borders of the same market forces that have operated for centuries at all levels of human economic activity—village markets, urban industries, or financial centers."[2]

(PowerPoint slide No. 17:)

"There are countless indicators that illustrate how goods, capital, and people, have become more globalized.

- *The value of trade (goods and services) as a percentage of world GDP increased from 42.1 percent in 1980 to 62.1 percent in 2007.*

- *Foreign direct investment increased from 6.5 percent of world GDP in 1980 to 31.8 percent in 2006.*

- *The stock of international claims (primarily bank loans), as a percentage of world GDP, increased from roughly 10 percent in 1980 to 48 percent in 2006.[1]*

- *The number of minutes spent on cross-border telephone calls, on a per-capita basis, increased from 7.3 in 1991 to 28.8 in 2006.[2]*

- *The number of foreign workers has increased from 78 million people (2.4 percent of the world population) in 1965 to 191 million people (3.0 percent of the world population) in 2005."[2]*

--I have a question, please – A young man standing by one of the doors to the conference room raised his right arm.

--Please go ahead! –Encouraged Xabier, with a smile.

--Yes, those are positive things about the process of globalization by the **International Monetary Fund (IMF)** and many other governments and organizations, but there are also many negative aspects which are beginning to show up in the communications media.

--Please go ahead, let us know about those negative aspects, of course.

--For one, globalization has contributed in many ways to the economies of developing countries where they now have jobs, better food supplies, and more international investment, on the positive side, to be fair. But globalization has also had very negative aspects on many small towns and cities in the Western world, including the

207

USA, Germany, France, Italy, and many other countries in the European Union (EU), to mention a few. To share a specific example, I will mention what has been happening in the Mid-West in the USA, where many small towns and large cities have been losing businesses and corporations involved in a variety of manufacturing enterprises in clothing, auto parts production, television set manufacturing and repair, domestic and kitchen product manufacturing, cell phone research and development, and many other products. Our companies simply closed their manufacturing facilities in those towns and cities and invested in other countries with lower labor costs. The simply moved jobs away from our towns and cities to other countries with lower labor costs. This situation, in fact, in in great part responsible for the unemployment in many of our towns and cities. This situation has also contributed to the unrest among citizens in those towns and cities in the Mid-West of the USA where they opted to vote for an individual like **Donald Trump** who promised them "**America first**" with better job employment opportunities. This situation is also happening in France with the candidate **Marine Le Pen** of the *National Front* political party in her efforts to win the presidency of the country. This situation is also responsible in part for the "**Brexit**" of the *United Kingdom* from the *European Union* (EU). Anyway, I could go on mentioning many other countries where globalization has had very negative impacts, in my opinion.

--Thank you for your examples of places and people where globalization may be having a negative contribution. Much appreciated, really. And now, we are going to ask my partner, *Xabier Elurmendi*, to join us and to guide us through the second half of this presentation. Xabier?

A Multi-Polar World

--Thank you, Kathy! My part on this second half of the presentation is to help us become aware of the fact that our world revolves around several main countries which we are going to call "poles", and as such we live in a *"multi-polar world"*.[3] For example, right after World War II we basically had to main poles: The United States of America (USA) and the Soviet Union, but also the USA and the Republic of China, two main powers today, as shown on *Figure 7.*

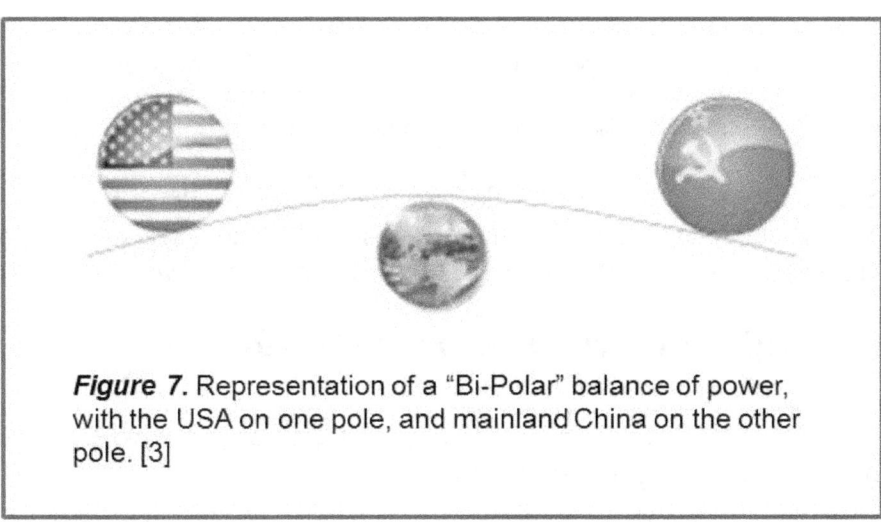

Figure 7. Representation of a "Bi-Polar" balance of power, with the USA on one pole, and mainland China on the other pole. [3]

Seventy years later, our balance of power has changed dramatically and we can say that we live in a multi-polar world represented by the USA, China, and other emerging countries like Russia, Brazil, Turkey, and the European Union (EU), as shown on *Figure 8.*

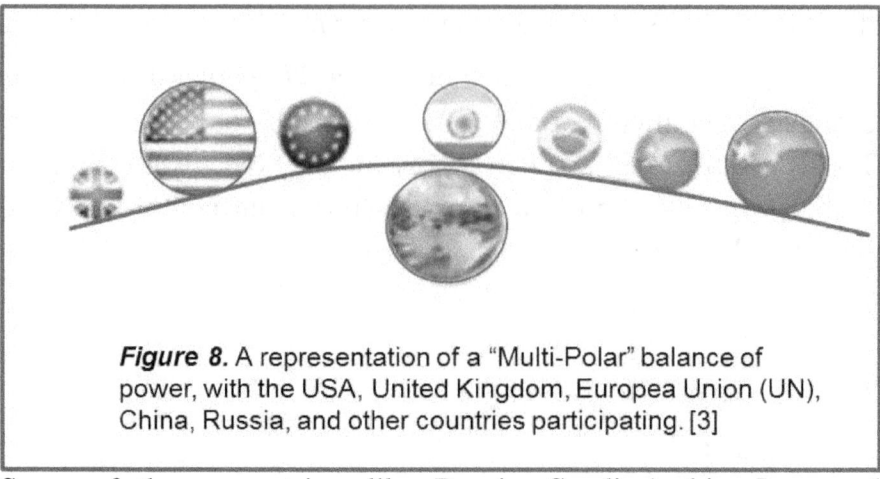

Figure 8. A representation of a "Multi-Polar" balance of power, with the USA, United Kingdom, Europea Union (UN), China, Russia, and other countries participating. [3]

Some of these countries, like Russia, Saudi Arabia, Iran, and Venezuela have large oil and gas resources and as such play a significant role in world politics and the balance of power.

A hypothetical New Order in the World

A hand is raised in the audience.

--So in this "new order" of the world, what is the role of the other not-so-strong, non-polar countries?

--That is an easy question with an easy answer. Those other countries, the great majority of the countries in our planet, "join" the large polar countries in economic, political, and military alliances and participate in a variety of ways, say, supplying low-cost labor, supplying agricultural lands as is the case of many countries in the African continent, and providing tourist sites.

--And what is the purpose of this Scenario 5, its role in considering the development of our global community and, in particular, its probability in the process of providing economic and political stability to our global community?

--I hear you, another good question, and for an answer we are going to ask **Dr. Eugene Finley**, the director of our IWP research team to step in with his comments and observations. Dr. Finley?

--Good afternoon to everyone in our conference room at the University of Arizona, here in Tucson, Arizona. As you already know, a main goal of this **Scenario 5** is to study and consider the processes involved in the development of an economic and political balance in our planet, in our global community. The thinking being that if we achieve a measure of economic and political balance in the planet, it would be more likely to continue as a species without wars, including nuclear warfare, which could mean the end of our species as we know it today. *The probability of achieving such economic and political balance?* Still very small, in the order of 10%-15% I would say, given our human nature as primates that we are, always wanting to have more of everything.

Still, considering such a *Scenario 5* would be positive in the sense that it would reduce the social, economic, and political inequalities in our world and, as such, would help us continue into the future. *For how many more years into the future?* We don't know, of course not, given that our history shows that as a human species we like to become involved in wars during periods of 5-10 years until we destroy several of our communities, decimate our populations, and then we take another 50-70 years of "peace" to recuperate, rebuild our cities, the structure of our societies, and once we reach a new period of affluence we can't help thinking of more wars. *It's a vicious cycle of life and death for our species.* A main problem is that nowadays we have thousands of nuclear warheads spread all over the planet, and a nuclear war would have catastrophic consequences for our species as *Homo Sapiens Sapiens.*

--Any other question on this *Scenario 5?* Good! Still, please remember to complete the one-page surveys available on your tables with your comments and even additional questions.

Again, thank you all for being here today. See you all next week as we continue with our Scenario Series.

Chapter 11:
Scenario 7: Search, Find, and Populate other Planets

*"These are the missing practical arguments: safeguarding the Earth from otherwise inevitable catastrophic impacts and hedging our bets on the many other threats, known and unknown, to the environment that sustains us. Without these arguments, a compelling case for sending humans to **Mars and elsewhere** might be lacking. But with them the buttressing arguments involving science, education, perspective, and hope I think a strong case can be made. If our long-term survival is at stake, we have **a basic responsibility to our species to venture to other worlds**."*
— **Carl Sagan**, "Pale Blue Dot: A Vision of the Human Future in Space", 1994

*"Unless we are willing to settle down into a world that is our prison, **we must be ready to move beyond Earth**... . People who view*

industrialization as a source of the Earth's troubles, its pollution, and the desecration of its surface, can only advocate that we give it up. This is something that we can't do; we have the tiger by the tail. We have 4.5 billion people on Earth. We can't support that many unless we're industrialized and technologically advanced. So, the idea is not to get rid of industrialization but to move it somewhere else. If we can move it a few thousand miles into space, we still have it, but not on Earth. Earth can then become a world of parks, farms, and wilderness without giving up the benefits of industrialization."
— **Isaac Asimov**, 'Our Future in the Cosmos — Space,' lecture given at the College of William and Mary, <u>full transcript online</u>, 1983.

--Good afternoon and welcome to our Seminar Series at the University of Arizona, here in Tucson, Arizona, USA. My name is **Kathy Thompson**, a member of the IWP research team, that is, the *International World Organization for Peace*, under the direction of **Dr. Eugene Finley**! And, yes, the second half of the presentation will be carried out by my partner, Xabier Elurmendi, also a member of the IWP team.

This afternoon we are going to do **Scenario 7**, with title:

"Search, Find, and Populate other Planets"

Some of you, here present, already knew that this scenario was in the making, as the IWP gathered the information needed from several sources, including the **Jet Propulsion Laboratory (JPL)**, and the **National Aeronautics and Space Administration (NASA)**. Yes, the IWP knows that similar scenarios have been presented on a number of movies and TV programs over the years, but this time we bring to you information gathered in the last few months, as well.

On 4 October 1957 the *Soviet Union* launched into a low Earth Orbit the first artificial Earth satellite, **Sputnik**, thus giving way to a gigantic "space race" between Russia and the USA. Nine years later, the American **Neil Armstrong** became the first person to make a space flight, and three years later, in 1966, "Armstrong's second and last spaceflight was as mission commander of the Apollo 11 Moon landing, in July 1969; on this mission, *Armstrong* and *Buzz Aldrin* descended to the lunar surface and spent two and a half hours exploring, while *Michael Collins* remained in lunar orbit in the Command Module." The race to space had begun. ***Should we as a species do interplanetary space travel, and should we try to place colonies in other planets, and which would be the reasons for doing so?*** On the following PowerPoint slides we listen to ideas and proposals for space travel and space colonization.

Interplanetary Space Travel
(PowerPoint Slide No. 1)

Are interstellar colonies possible? These would take the shape of *space vehicles* that move and exist between planets. On September 2014 a number of rocket scientists gathered in Houston's *George R. Brown Convention Center* to discuss—in level and serious tones—how to become a *space faring civilization*. The meeting was called the *100-Year Starship symposium*, and it's brought brains together once a year since 2011 to figure out what we need to do now if we want to have an interstellar space rocket a century from now. The group has made progress defining the challenges and pointing their noses toward solutions, but much work remains. Nonetheless, the 100-Year Starship adherents—backed by **NASA** and the **Defense Advanced Research Projects Agency (DARPA)**, keep plugging away. At their most recent gathering, several major hurdles emerged from their three days of discussion.

We would have to move fast. It's in the very definition of the project that we have to get to the stars. But it took Voyager 38 years just to get out of the solar system. Nobody has time for that. We have to figure out how to move fast through space. Engineers have a few ideas: Fusion rockets, ion drives, hydrogen-scooping ramjets, and antimatter annihilation systems, for instance.

215

Staying alive, staying alive. A starship has to be both sustainable and life-sustaining. It will be a closed ecosystem that must either have or produce everything humans need to survive and make more humans. Oxygen, food, and water are no-brainers. But we also need our micro biome, and we're only now starting to investigate what happens to our *symbiotic microbes* off Earth. The micro biomes of babies born in route remain a totally open question. And if they're in microgravity, will those babies' eye cells know where to migrate to become eye cells?

Crew-on-crew interaction. After a certain amount of time locked in a room together, you and even the most unobtrusive person will get on each other's nerves. You'll be like, "Why do you always have to look at me like that?" and they'll be like, "I'm not looking at you like anything." Etc. Now imagine that drawn out over years or decades on an interstellar mission. And beyond the everyday communication problems, disseminating information in a crisis—without causing panic or misinformation—could prove difficult. Imagine small-town social dynamics, in space. That's why it's important to choose a mix of crew members most likely to succeed. But first we have to find out what that mix is. Agencies across the world are already investigating this psychological puzzler. The *Mars HI-SEAS* experiment began its third mission on October 15, and six (lucky?) people will spend eight months on a simulated Martian colony in Hawaii. The *European Space Agency* teamed up from 2007-2011 with the *Russian Institute for Biomedical Problems* to do the Mars 500 mission, in which three different crews experienced isolation together aboard a pseudo-spaceship and then on a pseudo-Martian surface.

To become interstellar, we have to start with simply sending humans to space regularly, cheaply. We have to start manufacturing and mining off-planet. And we need to establish human colonies on the Moon, and probably on Mars, to make sure we have enough resources and practice before we set off toward *Earth 2.0.* Currently, the best bet rests in the hands of private industry. NASA's *asteroid capture mission* promises a step in that direction. But private companies like *Planetary Resources* and *Deep Space Industries*, which have both announced partnerships with NASA, may be able to make space travel and colonies their material domain faster and cheaper. [3]

Space Colonies

Are *space colonies* technologically possible, built *on the surface of some planets*, and what would be the implications?

Many arguments have been made for space colonization. The two most common are survival of human civilization and the biosphere in case of a planetary-scale disaster (natural or man-made), and the vast resources in space for expansion of human society. No space colonies have been built so far. Currently, the building of a space colony would present a set of huge challenges both technological and economic. Space settlements would have to provide for nearly all (or all) the material needs of hundreds or thousands of humans, in an environment out in space that is *very hostile to human life*. They would involve technologies, such as controlled ecological life support systems, that have yet to be developed in any meaningful way. They would also have to deal with the as yet unknown issue of how humans would behave and thrive in such places long-term. Because of the huge cost of sending anything from the surface of the Earth into orbit (roughly $20,000 USD per kilogram) a space colony would be a massively expensive project. There are no plans for building one by any large-scale organization, as shown on *Figure 1*, either government or private. However, there have been many proposals, speculations, and designs for space settlements that have been made, and there are a considerable number of space colonization advocates and groups.

Figure 1. A representation of a Space Colony in the near future. [1]

Obstacles and objections? Colonizing space would require massive amounts of financial, physical and human capital devoted to research, development, production, and deployment. Earth's natural resources do not increase to a noteworthy extent (which is in keeping with the "only one Earth" position of environmentalists). Thus, considerable efforts in colonizing places outside Earth would appear as a hazardous waste of the Earth's limited resources for an aim without a clear end. The fundamental problem of public things, needed for survival, such as space programs, is the "***free rider problem***". Convincing the public to fund such programs would require additional self-interest arguments: If the objective of space colonization is to provide a "***backup***" in case everyone on Earth is killed, then why should someone on Earth pay for something that is only useful after they are dead? This assumes that space colonization is ***not widely acknowledged as a sufficiently valuable social goal.*** Although seen as a relief to the problem of overpopulation, others have argued that space colonization is an impractical solution; in 1999, science fiction author *Arthur C. Clarke* said that "the population battle must be fought or won here on Earth".

Other objections include concern about creating a culture in which humans are no longer seen as human, but rather as material assets. The issues of *human dignity, morality, philosophy, culture, bioethics*, and the threat of megalomaniac leaders in these new "societies" would all have to be addressed in order for space colonization to meet the *psychological and social needs* of people living in isolated colonies.[4]

One hand is raised up in the air in the audience.

--Yes, please, what is your question?

--Miss Thompson, my question is a bit out of sequence, but wanted to ask you if you believe there is life out there, outside our planet Earth.

--It's OK, I can answer your question, it's an important question after all. A few years back I was a student. working a couple of summers at the *Jet Propulsion Laboratory (JPL)*, Pasadena, California, a federally funded research and development (R&D) center, as well as a *NASA* (*National Aeronautics and Space Administration*) field center. It was a great experience working with older and well experienced scientists, having the opportunity to work on space projects, while being able to compare the stories and claims made on TV by the news media and the general public with what with that being said by those experienced scientists at those R&D centers. On one hand *Carl Sagan*, the famous astrophysicist, would show up on the news media, on TV, talking about possible life in other planets, how possible it would be to travel to those planets, and how we needed then to send satellites and manned space vehicles to search and find life out there. At my work place, on the other hand, the older scientists would chuckle and say to themselves: *"Come on, Carl, we know there is no life out there!"* Really? So what was going on? The explanation is that administrators at JPL and NASA needed politicians and the general public to believe that there was life out there, life that could be found some day, in order for the US Congress to approve funding for those two large space agencies for the next year, and the next year, indefinitely. Simple. But that was my experience at JPL, nothing else.

--Three more PowerPoint slides with the following arguments.

(PowerPoint slides):

Argument A. Our *technology today*, including space technology, continues to advance rapidly today. Still, it would take several generations for space propulsion and transport technologies to emerge capable of taking groups of humans to some of those distant planets in a few months or a couple of years, and not the 30-50 years that would take today. The funding required by individual countries and international organizations would be enormous, quantities hard to justify in the light of very high priorities today for funding of programs in areas of population control, climate change, vanishing seafood resources, diminishing water supplies, drug trafficking, and wars over country boundaries and religious differences in the planet today. *Those needed space technologies would not become available* within the next 3-5 generations, a time period when attention to those problems here on Earth would have preference.

Argument B. Challenges on *human behavior and adaptation* to isolated colonies in planets and deep space would be considerable. If space technology is already highly complex, the challenges to human behavior and adaptation in isolated planets would be greater, in my opinion. Technology is something that we create in a few decades, a century or two, whereas human behavior and adaptability are traits that we have developed over Millions of years and, as such, very difficult to change in the matter of a few years, a few generations. *Differences over religious issues, population control, race, responsibility, and authority* would surface fairly soon, thus leading to community failure, and technological disaster via sabotage of life support systems. Said differently, until we humans learn to change and improve our human behavior here on Earth, experiments with space colonies in other planets would simply result in failures. Those space human settlements would be doomed from the first moment.

Argument C. Hope. Call it *hope or self-deception*? A few years, a few generations of exuberance, creativity, productivity, achievement, followed by decline and death, either on planet Earth or on some space colony out there. As a scientist I also would subscribe to the search for ways and means to create space colonies in the near future, even if the prospects are poor. As a species we need to exercise those three traits of *hope, creativity, and curiosity*, in our search for ways and means to survive, to keep on living as a species.

--Before going any further, could you give us a glimpse at the various themes that come with Scenario 7?

--Yes, of course, sorry! I forgot to do that at the very beginning of our presentation today. Here are two more PowerPoint slides with that information.

(PowerPoint slide):

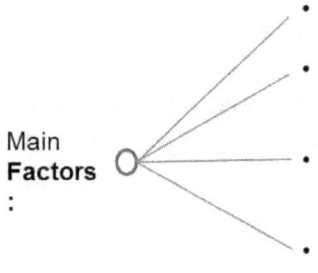

Main **Factors** :

- Currently the Earth is degrading environmentally at a fast rate.
- Our space technology would make it possible to build **space colonies** in other Planets.
- Our Human species is curious enough to want to explore other planets and create space colonies.
- There have been several "**close calls**" of **nuclear wars** on Earth, and the need exists to expand life to other planets.

Time Period: First **space colony on the Moon** could take place in the year 2050; other space colonies on other planets would then follow in the near future.

--Already we have the technology needed to create space colonies on the **Moon, Mars, and other nearby planet.**
--Ability of humans to co-exist in space colonies over decades is still unknown.
--Ability of humans to live on planets with much lower **gravity** is still unknown.
--Exchange of **energy and minerals** among space colonies may become possible.
--Cost of launching people and materials is still very e

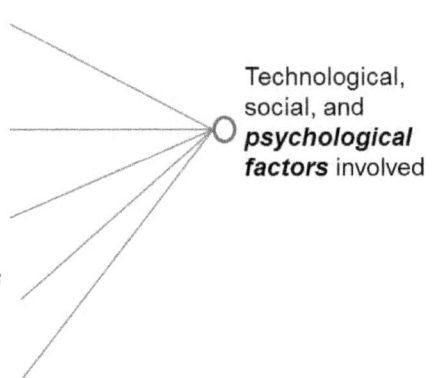

Technological, social, and **psychological factors** involved

Figure 2. Scenario 7, "**Search, Find, and Populate other Planets.**" (Part 1 of 2)

(PowerPoint slide):

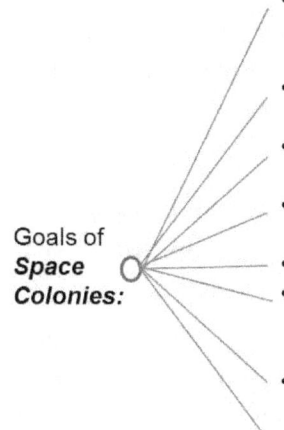

Goals of **Space Colonies:**

- To extend the **Life Cycle** of our Human species beyond the next few hundred or thousands of years.
- Whenever possible maintain our **ecological systems alive** and productive on Earth.
- Avoid the catastrophic, apocalyptic consquences of **nuclear wars** on our planet Earth.
- Conduct research to try and produce rocket engines **Faster-than-Light (FTL).**
- Alleviate **overpopulation** on Earth.
- Investigate **space commercialization** (e.g., solar power, asteroid mining, space manufacturing, other.)
- Initiate **space parterships** among Government and Private Sector corporations.
- Slow down the current dynamics of "**Political and military power balance**" on Earth, as to be able to direct energies and resources to space exploration and colonies.

--Some citizen communities may want to have the choice of staying on Earth or living on a space colony.
--Frequency of war among nations on Earth may be reduced, hopefully.
--Solar energy will be more accessible, reducing ecological degradation on Earth.
--Crime networks, however, may proliferate on Earth.

New **World Order**

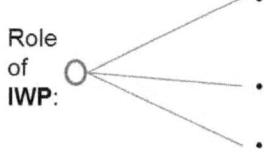

Role of **IWP:**

- Coordination of human and technological efforts among IWP, NASA, and other space agencies.
- Contribute "social models" to space colony initiatives.
- Coordination of peace and commercial treaties among nations on Earth.

Figure 3. Scenario 7, "**Search, Find, and Populate other Planets**." (Part 2 of 2)

Biosphere 2

Another hand rises in the air in the audience. A student from the *Department of Systems and Industrial Engineering (SIE)* of the ***University of Arizona (UofA).***

--Yes, your question, please.

--I understand there is a space research project somewhere in Arizona which has been initiated to study the feasibility of space colonies. What's the name and purpose of such project?

--Ahh, yes – answered Kathy – there is such a project in Arizona, but I am going to pass your question to my research partner, ***Xabier Elurmendi***, who recently visited that location and will continue with our presentation. Xabier?

--Thank you, Kathy! Yes, I will be glad to help with such question. As Kathy says, I visited with the people involved in that project by the name of ***Biosphere 2*** located in Oracle, Arizona, not too far away from our city of Tucson, really. Its mission is to do research in order to help us understand the feasibility of human life in space colonies and, in particular, the viability of closed ecological systems which would support and maintain human life in space colonies. It so happens that I have a couple of slides of this facility in this same computer, and it may take me only a couple of minutes to find those photographs of Biosphere 2, if I... There! I found the carpet in our computer, and there goes ***Figure 4*** onto the screen:

Figure 4. The **Biosphere-2 Project**, a research facility in Oracle, Arizona, USA, owned and directed by the **University of Arizona** (UofA) since 2011.

Its 5 biome areas constitute a 1,900 square meter *rainforest*, an 850 square meter *ocean with a coral reef*, a 450 square meter mangrove *wetlands*, a 1,300 square meter savannah *grassland*, a 1,400 square meter *fog desert*, a 2,500 square meter agricultural system, a human habitat, and a below-ground infrastructure. Power for heating and cooling of this diverse complex is supplied from the outside. Yes the project was initiated in in 1991, but the University of Arizona has assumed ownership of the project and has been directing its activities since 2011, only a few year ago.[6]

--And how is the Biosphere 2 project, this "space colony", working out?

--It has had its "up-and-downs", but it is working out quite well. Within the same carpet on my computer I have part of its story, let's take a look:

"The first closed mission lasted from September 26, 1991 to September 26, 1993. The crew were: medical doctor and researcher Roy Walford, Jane Poynter, Taber MacCallum,

225

Mark Nelson, Sally Silverstone, Abigail Alling, Mark Van Thillo, and Linda Leigh. The agricultural system produced 83% of the total diet, which included crops of bananas, papayas, sweet potatoes, beets, peanuts, lablab and cowpea beans, rice, and wheat. No toxic chemicals could be used, since they would impact health. During the first year the eight inhabitants reported continual hunger. During the second year, the crew produced over a ton more food, average caloric intake increased, and they regained some weight lost during the first year.

They consumed the same low-calorie, nutrient-dense diet which Roy Walford had studied in his research on extending lifespan through diet. *Medical markers indicated the health of the crew during the two years was excellent.* They showed the same improvement in health indices such as lowering of blood cholesterol, blood pressure, enhancement of immune system. They lost an average of 16% of their pre-entry body weight before stabilizing and regaining some weight during their second year. Subsequent studies showed that the biospherians' metabolism became more efficient at extracting nutrients from their food as an adaptation to the low-calorie, high nutrient diet.

Some of the domestic animals that were planned for the agricultural area during the first mission included four pygmy goats and one billy goat from the plateau region of Nigeria, 35 hens and three roosters (a mix of Indian jungle fowl (*Gallus gallus*), Japanese silky bantam, and a hybrid of these), two sows and one boar pig (feral), as well as tilapia fish grown in a rice and azolla pond system originating millennia ago in China.

A strategy of "species-packing" was practiced to ensure that food webs and ecological function could be maintained if some species did not survive. The fog desert area became more chaparral due to condensation from the space frame. The savannah was seasonally active; its biomass was cut and stored by the crew as part of their management of carbon dioxide. Rainforest pioneer species grew rapidly, but trees there and in the savannah suffered from etiolation and weakness caused by lack of stress wood, normally created in response to winds in natural conditions. Corals reproduced in the ocean area, and crew helped maintain ocean system health by hand-harvesting algae from the corals, manipulating calcium carbonate and pH levels to prevent the ocean becoming too acidic, and by installing an improved protein skimmer to supplement the algae turf scrubber system originally installed to remove excess nutrients.[6]

--Did anything not work well, any problems in the management or operation of this research facility?

--Yes, there were some problems, as the report continues:

On April 1, 1994 *a severe dispute within the management team* led to the ousting of the on-site management by federal marshals serving a restraining order, and financier Ed Bass hired Stephen Bannon, manager of the Bannon & Co. investment banking team from Beverly Hills, California, to run Space Biospheres Ventures. Some Biosphere-ites were concerned about Bannon, who had previously investigated cost overruns at the site. Two former Biosphere 2 crew members flew back to Arizona to protest the hire and broke into the compound to warn current crew members that Bannon and the new management would jeopardize their safety.

At 3 am on April 5, 1994, Abigail Alling and Mark Van Thillo, members of the first crew, allegedly vandalized the project from outside, opening one double-airlock door and three single door emergency exits, leaving them open for approximately fifteen minutes. Five panes of glass were also broken. Alling later told the Chicago Tribune that she "considered the Biosphere to be in an emergency state... In no way was it sabotage. It was my responsibility." About 10% of the biosphere's air was exchanged with the outside during this time, according to systems analyst Donella Meadows, who received a communication from Ms. Alling in which she explained that she and Van Thillo judged it their ethical duty to give those inside the choice of continuing with the drastically changed human experiment or leaving, as they

227

didn't know what the crew had been told of the new situation. "On April 1, 1994, at approximately 10 AM ... limousines arrived on the biosphere site ... with two investment bankers hired by Mr. Bass ... They arrived with a temporary restraining order to take over direct control of the project ... With them were *6-8 police officers* hired by the Bass organization ... They immediately changed locks on the offices ... All communication systems were changed (telephone and access codes), and [we] were prevented from receiving any data regarding safety, operations, and research of Biosphere 2." Alling emphasized several times in her letter that the "bankers" who suddenly took over "knew nothing technically or scientifically, and little about the biospherian crew."

Mission 2 was ended prematurely on September 6, 1994. No further total system science has emerged from Biosphere 2 since that date. Then, on June 26, 2007, the *University of Arizona* announced it would take over research at the *Biosphere 2*. The announcement ended fears that the structure would be demolished. University officials said private gifts and grants enabled them to cover research and operating costs for three years with the possibility of extending funding for ten years. It was extended for ten years, and is now engaged in research projects including research into the terrestrial water cycle and how it relates to ecology, atmospheric science, soil geochemistry, and climate change. In June 2011, the University announced that it would assume full ownership of Biosphere 2, effective July 1, same year.

Promotion of Space Colonies by the Private Sector

--Anything you have to share with our audience regarding comments by people in the *scientific community*?

--Yes. In fact, I have copy in the same carpet of an interview conducted by James Fallows, a reporter for The Atlantic magazine with questions he had for Eric C. Anderson, a young aerospace engineer (born 1974), and co-founder of Space Adventures Ltd., the first commercial space flight company which has arranged for eight missions for private citizens to the International Space Station since 2001. Here we have the *questions and answers*: [5]

(PowerPoint slides, the Interview):

James Fallows (JF): Space exploration seems to have lost its hold on the public imagination, compared with a generation ago.

Eric Anderson (EA): I think absolutely they are right to feel a little bit disappointed. On April 12, 1961, the first human being, Yuri Gagarin, goes to space. Then, July 29, 1969: We're on the moon. If you and I were doing this interview on July 30, 1969 and you had asked me what space exploration would be like in the year 2013, I would've told you it would be far more advanced than it is now.

So I think the reality is that space was unnaturally accelerated by this Cold War conflict between the United States and the Soviet Union during the 1960s. Then, in the early part of the '70s, that sort of slowed down. The latter half of the '70s brought terrible economic trouble in the U.S., which really set the space program way back. In the '80s, it was the reverse. The Soviets basically ran out of money and then the Soviet Union collapsed. Then in the '90s we were sort of figuring out how to re-set ourselves in a post-Soviet world. It was in the mid-'90s that commercial revenues in space started to eclipse government revenues—that was mainly for communication satellites and things like that.

So that part of the industry has gone pretty well. Every day we use GPS and DirecTV and get the weather, and that sort of stuff. But human flight has just been totally crimped. The number of people going to space, and the missions they were doing, went down. The Space Shuttle was so much over budget that it just was impossible for us to really do any real exploration. That's a long-winded answer, but yes: There's every reason for people to be disappointed with where we are now, particularly with regard to human space flight.

JF: Why should people be excited about what lies ahead?

EA: In the next generation or two—say the next 30 to 60 years—there will be an irreversible human migration to a permanent space colony. Some people will tell you that this new colony will be on the moon, or an asteroid—in my opinion asteroids are a great place to go, but mostly for mining. I think the location is likely to be Mars. This Mars colony will start off with a few thousand people, and then it may grow over 100 years to a few million people, but it will be there permanently. That should be really exciting, to be alive during that stage of humanity's history.

JF: I have to ask—really? This will really happen?

EA: I really do believe it will. First of all, the key to making it happen is to reduce the cost of transportation into space. My colleague Elon Musk is aiming to get the cost of a flight to Mars down to half a million dollars a person. I think that even if it costs maybe a few million dollars a person to launch to Mars, a colony could be feasible. To me the question is, does it happen in the next 30 years, or does it happen in the next 60 to 70 years? There's no question it's going to happen in this century, and that's a pretty exciting thing.

JF: Apart from the cost of transport, what are the challenges in making that a reality? Are they cost and engineering challenges, or are they basic science problems?

EA: I think it's all about the economics. There is no technological or engineering challenge.

One key to making all this happen is that we need to use the resources of space to help us colonize space. It would have been pretty tough for the settlers who went to California if they'd had to bring every supply they would ever need along with them from the East Coast.

That's why Planetary Resources exists. The near-Earth asteroids, which are very, very close to the Earth, are filled with resources that would be useful for people wanting to go to Mars, or anywhere else in the solar system. They contain precious resources like water, rocket fuel, strategic metals, as *Figure 2* portrays. So first there needs to be a reduction in the cost of getting off the Earth's surface, and then there needs to be the ability to "live off the land" by using the resources in space.

Figure 5. A representation of a mining operation on a Space Colony. [1]

JF: Again—really? To the general public, asteroid mining just has a fantastic-slash-wacky connotation. How practical is this?

EA: When [co-founder] Peter Diamandis and I conceived of the company, we knew it would be a multi-decade effort. From history, we knew that frontiers are opened by access to resources. We would like to see a future where humans are expanding the sphere of influence of humanity into space.

To make asteroid mining viable, we need spacecraft that can launch and operate in space considerably less expensively than has traditionally been the case. If we are able to do that, then asteroid mining can be profitable—very much so. When you ask "Is it viable?," I'll be the first one to tell you how risky this proposition is, and how there is a significant possibility that we could fail in a particular mission or technology, or fall short of our goals.

But we have found ways to reduce the cost of space exploration already. For example, our prospecting mission to a set of targeted asteroids will use the Arkyd line of spacecraft. The first of that series, the Arkyd-100, would have cost $100 million, minimum, in the traditional aerospace way of business and operation. But with the engineering talent we have, and by using commercially available parts and allowing ourselves to take appropriate risks, we've been able to bring that cost down to $4 or $5 million dollars.

In 10 years or so, what we'd really like to do is get robotic exploration of space in line with Moore's Law [the tech-world maxim that the price for computing power falls by half every 18 months]. Remember, asteroid mining doesn't involve people. We want to transition space exploration from a linear technology into an exponential one, and create an industry that can flourish off of exponential technologies such as artificial intelligence and machine learning.

Our first missions, for asteroid reconnaissance, will be launching in the next two to three years. For these missions, we're going to launch small swarms of spacecraft. When I say small, I mean we'll send three or four spacecraft, and each one of those spacecraft may weigh only 30 pounds. But they will have optical sensors that are better than any camera available today. They will send back imagery, they'll map the gravity field, they'll use telescopic remote sensing and spectroscopy to tell us exactly what materials are in the asteroid. It will be possible to know more about an ore body that's 10 million miles away from us in space than it would be to know about an ore body 10 miles below the Earth's surface.

We're really not talking about if; we're talking about when.

JF: Apart from the practicalities of asteroid mining, what is it going to mean in spiritual and philosophical ways for people to leave the Earth? I guess this is taking us back to the science fiction of the '50s and '60s, but what do you think?

EA: I've thought a lot about that. The interesting thing will be to see why the people who go to Mars, or to a colony on the moon, or to an asteroid, decide to go there. Will they go there because they're escaping something? Will they go there because they're curious? Will they go to make money?

Throughout history, most of the frontiers that we have had on the Earth have been opened up because people were seeking land—new hunting grounds, or fertile locations for cattle—or mining for gold or precious metals. But occasionally they would go somewhere new because they were seeking religious freedom or some other kind of freedom.

So I don't actually know why people will go. Will the Earth be so ravaged by war, or catastrophic climate change, or whatever else, that people will want to leave?

JF: In addition to the forces you mentioned, over the last half millennium or more, the search for new territory has been powerfully driven by national rivalries. The French, the English, the Spanish and others were seeking new territory in which to spread their influence. Do you imagine the national rivalries on Earth being soothed by space exploration? Or rather being aggravated by space exploration, the way the exploration of the New World was?

EA: I think it's an excellent question, and I think it's inevitable. The Outer Space Treaty, which was signed in 1967, basically says that no nation can claim a celestial body for its own sovereignty. And it also says that anything that is launched from a particular nation, that nation is responsible for, if it crashes into another nation or something like that. But I don't see the Outer Space Treaty living another 100 years.

I think that history repeats itself, and all the same things that happened in our history over the last thousand years will happen in one form or another in the next thousand years. Nowadays things are accelerated, it won't take as long for those cycles of history to happen—because we have faster means of communication, faster democracies, faster governments. The consequences of action, of economic and political and social drivers, can be felt and reacted to faster than they have been in the past.

But those same things will happen. If the first colonists going to Mars are all American, what kind of system do you think they're going to want to set up on Mars? And how are other countries going to feel about that? And at what point will the Americans just pull out of the Outer Space Treaty? Or maybe it'll be the Chinese—the Chinese could get to Mars long before us. Who knows? But being there is 99 percent of it and I think that when the dam breaks and it's possible to travel at a reasonable cost in space outside the Earth's very-near vicinity, all sorts of things are going to change.

And one of the other tenets of the Outer Space Treaty is that space will not be weaponized. I hope that lasts for a long, long, long time, but I mean, who knows, it seems like a pipe dream to think that would last forever.

JF: About the environment: Are you thinking space could be not just an escape from a ravaged Earth but a way to save the Earth?

EA: There's a huge environmental cost to mining on Earth. But there are lots of strategic materials and metals that we can get in space and that will be necessary for us if we want to create abundance and prosperity generations from now on Earth. We sort of had a freebie over the past couple hundred years—we figured out that you can burn coal and fossil fuels and give all the economies of the world a big boost. But that's about to end. Not only do we have to transition to a new form of energy, we also have to transition to a new form of resources. And the resources of the nearest asteroids make the resources on Earth pale by comparison. There are enough resources in the nearest asteroids to support human society and civilization for thousands of years.

I'm not suggesting that we're going to start using resources from space next year. But over the next 20 years, resources in space will most likely be used to explore our solar system. And eventually we'll start bringing them back to Earth. Wouldn't it be great if one day, all of the heavy industries of the Earth—mining and energy production and manufacturing—were done somewhere else, and the Earth could be used for living, keeping it as it should be, which is a bright-blue planet with lots of green?

JF: Here's my last question. When I was a kid in the Baby Boom era, there was a genuine national excitement about space. Do you think that mood in the United States needs to be recreated for the populace as a whole? With an overall national excitement or sense of mission about space exploration, like in the 1960s? Or, on the contrary, is this something that should and can be left to people who see a business or scientific opportunity?

EA: If you look at polls, about half the population says that if it were at a price they could afford, and it were safe, they would go to space themselves. They would love to see the Earth from space. I don't know what that means in terms of gauging support. But clearly the more people are interested in and supportive of space exploration, the faster the industry will grow.

I think spending a half a percent of GDP on space, on space exploration, would be a very wise investment, whether that investment comes from the government itself or from just private industry. There are few things that inspire human engineering, human ingenuity, and the human spirit more than space exploration. Kids love space, and they love dinosaurs, and they love all those fantastical things that can happen when you push the boundaries. It's the same reason that, when my little one crawls out of her crib at night, she peeks around the corner to see what's there. This is curiosity.

We have enough perspective on ourselves and the universe to know that we just inhabit this tiny little corner of the universe. Humans are curious; so to say that we're not interested in space would put us [at odds with] the very core of our being as humans, in a world where we've defined a limit that we can never go beyond.

We obviously have huge problems on Earth, and nobody's saying that we should try to go develop space in lieu of solving our problems on Earth. But the fact of the matter is that we should always be doing things that inspire our youth and ourselves, and try to bring out the best parts of human nature."[5]

Seven New Planets discovered

Another hand is raised in the audience, a reporter from the *Tucson Daily* newspaper in Tucson City.

--Yes, it is a very interesting interview, for sure. However, *Mars* may still be a very difficult planet to look into with the possibility of creating a space colony there. Do we know of any other potential sites, other planets or large asteroids, which can be thought as sites for space colonies in the near future?

--A very good question, and we are lucky to learn that as recently as February 2017 a total of 7 Planets have been found orbiting a small star by the name of TRAPPIST-1. Here is part of a report on such finding:

(PowerPoint slides):

"Astronomers have found at least seven Earth-sized planets orbiting the same star *40 light-years away*, according to a study published Wednesday in the journal Nature. The findings were also announced at a news conference at NASA Headquarters in Washington.

This discovery outside of our solar system is *rare* because the planets have the winning combination of being similar in size to Earth and being all temperate, meaning they could have water on their surfaces and potentially support life.

"This is the first time that so many planets of this kind are found around the same star," said *Michaël Gillon*, lead study author and astronomer at the *University of Liège in Belgiu*m.

The seven exoplanets were all found in tight formation around an ultra cool dwarf star called *TRAPPIST-1*, as shown on *Figure 3*. Estimates of their mass also indicate that they are rocky planets, rather than being gaseous like Jupiter. Three planets are in the habitable zone of the star, known as

TRAPPIST-1e, f and g, and may even have oceans on the surface.[2]

Figure 6. A representation of the Trappist-1 star with seven Earth-like planets orbiting it. [2]

The TRAPPIST-1 star, an ultracool dwarf, has seven Earth-size planets orbiting it.

The researchers believe that TRAPPIST-1f in particular is the best candidate for supporting life. It's a bit cooler than Earth, but could be suitable with the right atmosphere and enough greenhouse gases.

If TRAPPIST-1 sounds familiar, that's because these researchers announced the discovery of three initial planets orbiting the same star in May. The new research increased that number to seven planets total.

"I think we've made a crucial step towards finding if there is life out there," said *Amaury Triaud*, one of the study authors and an astronomer at the *University of Cambridge*. "I don't think any time before we had the right planets to discover and find out if there was (life). Here, if life managed to thrive and releases gases similar to what we have on Earth, we will know."

Life may begin and evolve differently on other planets, so finding the gases that indicate life is key, the researchers added.

"This discovery could be a significant piece in the puzzle of finding habitable environments, places that are conducive to life," said Thomas Zurbuchen, associate administrator of NASA's Science Mission Directorate. "Answering the question 'are we alone?' is a top science priority, and finding so many planets like these for the first time in the habitable zone is a remarkable step forward toward that goal."

And as we've learned from studying and discovering exoplanets before, where there is one, there are more, said **Sara Seager**, professor of planetary science and physics at *Massachusetts Institute of Technology* (**MIT**). Seager and other researchers are encouraged by the discovery of this system because it improves our chances of finding another habitable planet, like Earth, in the future, by knowing where to look.

--Yes, it is an amazing discovery, seven planets which have some similarities to our planet Earth, but the fact that they are **40 light-years** away from Earth brings down my enthusiasm. Also, in your opinion, what is the **probability** of space colonies happening and when may we see the first space colony in our galaxy?

--The experts believe that such a first space colony may take place on the Moon by the year 2050, but it may be of an exploratory nature as there are still many technological, social, and psychological challenges to address in detail. Are there any other questions? – Xabier looked out into the audience during a few seconds – Very well, then please go ahead and fill the one-page survey which is to be found on your tables.

Again, our thanks to everyone here today for your curiosity, questions, and great comments.

Thank you, all!

***

Chapter 12:
Scenario 8: Coming to terms with our Human Nature and Place in the Universe

"Corruption, embezzlement, fraud, these are all characteristics which exist everywhere. It is regrettably the way human nature functions, whether we like it or not. What successful economies do is keep it to a minimum. No one has ever eliminated any of that stuff."
--Alan Greenspan
Read more at:
https://www.brainyquote.com/quotes/keywords/human_nature.html

"If you look at life with any honesty and intelligence, it's clear that human nature is dark, vile, selfish, and despondent. But I also see a force in human nature, namely grace, that sometimes works against our natural moral entropy."

--Come in, come in, please. Find a seat and make yourselves comfortable. Welcome to our Seminar Series at the *University of Arizona*, here in Tucson, Arizona, USA. My name is *Xabier Elurmendi*, and it is my turn to do the first half of this presentation with title:

"Coming to terms with our Human Nature and Place in the Universe"

As you already know this is *Scenario 8*, the last one in our Seminar Series under the direction of *Dr. Eugene Finley* of the *International World Organization for Peace (IWP).* The second half of this presentation will be conducted by my partner in the IWP research team, *Miss Kathy Thompson*, who is very familiar with the topics we are going to share with you, as listed in this PowerPoint slide:

Contents:
- **Understand human origins 50,000 years ago in Africa**
- **Recognize deceit, fraud, murder, piracy, and wars in human nature**
- **We are alone in the Universe, a unique gift**
- **There is no heaven, no hell, and no afterlife**
- **The need for recognition in our social hierarchies.**

Are Humans about to become Extinct?

Yes, this scenario is probably going to be one of the most difficult to communicate and to share with you, as it deals with many attributes of our own human nature, as well as with facts about our species in the Universe, many of which we have chosen to either ignore in the development of our civilizations, or we simply had no knowledge of those attributes.

Some of the scientists that we introduce here are also very profound and radical because of their beliefs which claim that our humanity, our species, is bound for extinction, that it is too late to try to redirect matters to a longer existence in our planet. In fact to try and redirect matters we would have to address those five statements above and consider modifying our current beliefs, not an easy matter. *Our objective in this Scenario 8 is precisely that, to consider whether our current beliefs could be modified in an effort to extend our life cycle on our planet Earth.*

Is everyone then ready to look into the details of those 5 statements? – A few seconds go by – Great! Then we continue with a report on Dr. Frank Fenner (1914-2010), an evolutionist biologist, and an Emeritus professor from the *Australian National University*:[1]

"Professor *Frank Fenner* says humanity is finished. It's already too late to save ourselves from the suicidal future we've created where the ecosystem can no longer support human life. *"We're going to become extinct,"* the scientist says. *"Whatever we do now is too late."*

Frank Fenner, now passed away, was a fellow of the *Australian Academy of Science and of the Royal Society*. He's published "hundreds of scientific papers and written or co-written 22 books," says *The Australian*, which prints these words from an interview with Fenner:

We'll undergo the same fate as the people on Easter Island... Climate change is just at the very beginning ... The human species is likely to go the same way as many of the species that we've seen disappear. Homo sapiens will become extinct, perhaps within 100 years ... It's an irreversible situation. I think it's too late. I try not to express that because people are trying to do something, but they keep putting it off.

The really scary thing is not that Fenner believes humanity is headed for extinction. Rather, the bigger threat to us all is **what the globalists might attempt to do in an effort to change   course**.

Hint: It involves a planet with six billion fewer people than live on it today.

Too many people, not enough food, water and other resources

Fenner's argument is founded on the idea that **the population is already too large to stop the mass extinction**. It is "unbridled consumption" that's going to lead humanity to inadvertent mass suicide, he says.

Fenner is no Sarah Connor, however. He's not a psych ward patient; he's an esteemed scientist and a world-renowned expert in pox viruses. On the extinction of the human species, he explains "Mitigation would slow things down a bit, but there are too many people here already."[1]

The collision of consumption and depletion. What we're all really witnessing here is the violent collision of two global trends: consumption and depletion.

Consumption is what the corporations want everybody to increase. They want more people to buy more stuff, throw away more stuff and replace it with yet even more stuff, some of which you stuff into your own face so that you become diabetic and need to buy more medical stuff. Corporations only profit when people consume, after all, so almost every message that's pushed in our modern society is engineered around an agenda to promote mass consumption for the purpose of boosting corporate profits. (Regardless of what happens to the environment as a result.)

Depletion is what happens when the planet's citizens use up all the finite resources: hydrocarbons (fossil fuels), fresh water (aquifers), topsoil, clear-cutting forests for GMO soybean production, strip-mining rare earth metals to build guilt-supported wind farms for "green" energy, destroying natural marine ecosystems, and so on.

When consumption collides with depletion, guess what happens? *Extinction*. Easter Island 2.0, in other words, but on a planetary scale.

There is no such thing as unlimited growth that's also sustainable. Nearly all the economic models underpinning human civilization are based on the fraudulent concept of **limitless growth and expansion**. But the very idea is a fraud. Our planet is finite, obviously, or else it would occupy the entire universe (and then some).

Because the Earth is finite, it's resources are also finite at any given moment. While things like fossil fuels and freshwater aquifers can be recharged over time, that time scale vastly exceeds the relatively short window of human time in which they are being exhausted. (A given water aquifer, for example, might be pumped dry by human agriculture in 50 years, while it could take 500 years to recharge.)

The real question is ***where are we now on the consumption vs. depletion curve?*** Nearly everyone in business and industry says there's nothing to worry about. Keep consuming! As Jay Leno used to say when pimping for Doritos, "Eat all you want, we'll make more!"

That's all great until the day the water runs out, the crops shrivel up, the pollinators vanish and the food supply implodes. "As the population keeps growing to seven, eight or nine billion, there will be a lot more wars over food," says Professor Fenner. Along with those wars, we'll also see wars over water, honeybees, seeds and perhaps even atmospheric oxygen.

These wars, Fenner seems convinced, will ultimately be fruitless because even the victors are doomed as the ecological destruction set in motion by shortsighted human beings plays out, devastating the planet's ability to sustain human life.

When the ecosystem collapses, after all, the collapse of human life must follow.

The reason Fenner can utter these things, by the way, is because he's long retired. Nobody can threaten to derail his academic career or halt his pension. He's not vying for a Nobel Prize or a coveted professorship at *Politically Correct University*. So he's uttering what he sees as truthful yet dire: Humanity is already doomed, and "green living" is a cruel joke that won't make any difference at all in the long run, he's convinced.

Let's hope Fenner is wrong. But even if he is, that doesn't mean the power-hungry globalists running our world won't try to murder six billion people anyway... especially if it means, in their minds, "saving humanity's future from ecological collapse."

Michelle Obama, by the way, has just announced that America's school lunch program will start serving **free soylent green burgers to everyone!**

Solutions? Probably not any that you're willing to actually pursue. With all this in mind, what can we do to help reverse humanity's apparent course towards self-extinction?

Forget about all the wimpy "green living tips" you'll find in dumbed-down newsstand magazines. Sorting your trash into three different recycling bins won't make a lick of difference in the long run. And no, your government-mandated low-flush toilet doesn't make you "environmentally conscious." The very fact that you're buying fruit in the winter that's shipped from South America is a total contradiction of core environmental principles.

If you really want to contribute in a meaningful way to a sustainable world, you need to *stop driving a vehicle, stop buying food at the grocery store, abandon the very idea of a green lawn, grow most of your own food and live in a tiny mud hut with no electricity or running water.* Eat seasonally-grown foods grown no more than 250 miles from your home, in other words. Otherwise, even the way you eat is a daily assault on the planet.

Have I achieved all this? Not by a long shot. I still eat fresh organic produce grown in Chile. I still drive a gas-guzzling vehicle. I still use electricity produced by coal. But I'm way ahead of most on several fronts, including living 100% on my own rainwater, growing an increasing percentage of my own food, raising my own free-range chickens and experimenting with gravity-fed irrigation systems for long-term food sustainability."[1]

A hand is raised in the audience.

--Surely there must be *other voices* who claim that there is still hope for humanity. Can you mention some of those voices?

--Certainly, that is the case. There are other scientists, like *James Lovelock* (1919), who believe that there is hope for mankind in spite of all the mistakes already committed by our species. He lives in Devon, England, and he is best known for proposing the *Gaia hypothesis* which postulates that the Earth functions as a self-regulating system:[2]

> "A decade ago, he predicted that billions would be wiped out by floods, drought and famine by 2040. He is more circumspect about that date these days, but he has not changed his underlying belief that the consequences of global warming will catch up with us eventually. His conviction that humans are incapable of reversing them – and that it is in any case too late to try – is also unaltered. In the week when the *Grantham Research Institute on Climate Change* reported that the world is still miles off meeting its 2030 carbon emission targets, Lovelock cannot easily be dismissed.

> There are other doomsayers. What makes this one so unusual is his confounding cheerfulness about the approaching apocalypse. His optimism rests on his faith in Gaia – his revolutionary theory, first formulated in the 1970s, that our

planet is not just a rock but a complex, self-regulating organism geared to the long-term sustenance of life. This means, among other things, that if there are too many people for the Earth to support, *Gaia – Earth* – will find a way to get rid of the excess, and carry on.

Lovelock's concern is less with the survival of humanity than with the continuation of life itself. Against that imperative, the decimation of nations is almost inconsequential to him. "You know, I look with a great deal of equanimity on some sort of happening – not too rapid – that reduces our population down to about a billion," he says, five minutes into our meeting. "*I think the Earth would be happier ... A population in England of five or 10 million? Yes, I think that sounds about right.*" To him, even the prospect of nuclear holocaust has its upside. "*The civilizations of the northern hemisphere would be utterly destroyed, no doubt about it,*" he says, "*but it would give life elsewhere a chance to recover. I think actually that Gaia might heave a sigh of relief.*"

As a man of science, he remains agnostic on the subject of God. And yet, he says, "I am beginning to swing round, to think more and more, that there's something in Barrow and Tipler's cosmic-anthropic principle – the idea that the universe was set up in such a way that the formation of intelligent life on some planet somewhere was inevitable ... The more you look at the universe, the more puzzling it is that all the figures are just right for the appearance on this planet of people like us."

For the time being our species may be, as he has written, "scared and confused, like a colony of red ants exposed when we lift the garden slab that is the lid of their nest". But he is also content to be one of those ants, because he sees a kind of beauty in that confusion – and perhaps even some sort of grand design. "Humanity may be as important to Earth, to Gaia, as the first photo-synthesizers," he thinks. "We are the first species to harvest information ... that is something very special."

Above all he is convinced that mankind can recover itself – and in this he may be a product of his vanishing generation. Some years ago, at a lecture in Edinburgh, I heard him reminisce how marvelously the British nation had pulled together when threatened by Nazi invasion, but that it had taken that existential threat to make them do so. *When the climate crisis finally breaks, he believes, the world's differences will again be put aside – and our species, for all its present idiocies, will pull together in a way that will astonish the cynics among us.*"[2]

Understand human origins 50,000 years ago in Africa

--As you may recall, we already took a look in *Scenario 5* at the origins of our species, coming out of Africa some 50,000 years ago.

One more time, all races in the planet originated with Homo Sapiens Sapiens, that group of hominids made up of a few dozens of families that left Africa (area of Ethiopia) 50,000 years ago, spreading and reaching all five continents. Greeks, Tibetans, Galicians, Germans, Catalans, Russians, Native Americans, *Basques*, Rumanians, Jews, Spaniards, Irish, Palestinians, Moais... all. We all have our roots in Africa, we all come from Africa. *Our ancestors were all of dark skin, black and curly hair. Let us review our beliefs and perspectives in order to reflect this incredible and marvelous scientific reality*, I propose.

--Yes, your question, please.

--It's just a comment. I appreciate the details and the graphs, but there are still many people in our communities whom would not be very kind to those slides and their graphs. I mean those many people have been brought up in within religious families and communities, and it would be very difficult and hard for them to set those beliefs aside, for one thing.

--True, very true. But that is precisely what we are trying to communicate in this Scenario 8, to bring out the scientific findings of the last 50-75 years. I say these things because we, as a human species, would be able to stay away from wars if we understood

249

well that we are all "cousins", we are all the same family, no matter what ethnic group we come from, hopefully.

Recognize deceit, fraud, murder, piracy, and wars in human nature

American scientist **Robert Trivers**[3] and others have presented arguments that attest that *"deceit and lying is an advantage in the evolution process."* The chameleon *changes color* in the tropical forests in order to blend with the surrounding trees and leaves, thus tricking potential larger animals to escape from being eaten, or deceiving victims to be eaten. A *survival strategy*. No *"morality"* is involved. Army A is about to attack soldiers and people inside a city, but tries to deceive those city dwellers into thinking that the attack will come *from the north*, when in reality the attack will come from the south. The lion will approach the gazelle slowly through the tall grass, without being seen, without making noise, without giving away his true intentions. And so forth. Those of us alive today, whether human beings or any other animal, are the descendants of animal groups that were successful at deceiving other predators over thousands of years.

Scary proposition? Possibly. Looking at "the other side of the coin", it can be argued that those human ethnic groups and society models that are slow in the "deceiving process" may be annihilated, replaced by other groups in the future. An issue of survival strategy, not of "morality."

There is no heaven, no hell, no afterlife

A lady standing by the main door to the conference room raises her hand.

--Yes, please your question.

--At the beginning of this presentation you said something about "no heaven, no hell, and no afterlife", which I think contradicts the contents of an earlier Scenario -- Scenario 5 it was, I believe-- and in it your group talked about "Diversity of Religions and Beliefs in the World." Doesn't that statement contradict in some way the contents of that earlier Scenario?

--Well, yes and no. In that earlier scenario, *Scenario 5*, we wanted to address the possibility of co-existence of a large variety of religions in our communities, in our planet Earth. In this **Scenario 8** our IWP research team would like to go a bit further, and for that purpose I am going to ask my partner in the team, *Kathy Thompson,* to step in and guide us through this process in answer to your question.

--Thank you, Xabier! This time I would like to share with you a set of tenets which many folks in our IWP research team ascribe to. I am saying that because they are scientists you should believe in them and, instead, simply to consider them in your mind over the next months and years. Talk about them with friends and professionals in your respective communities and you may be surprised at the responses you get, the findings shared by many folks. Again, what we are saying here is that if we ever got to realize that this is a very special opportunity, a unique opportunity to live our lives, one time and never again, we just might be able to appreciate life much more, cherish it, and protect it, away from fratricide wars. Here are a couple of PowerPoint slides:

(PowerPoint slide 19)

We are about to complete this Seminar Series, so many of us in the IWP research team would like to emphasize one more time our views on a brand of Atheism that we like to call **Community-based Atheism (CBA),** one that pretends to go beyond mainstream Atheism. As such, it goes beyond the assertions on the non-existence of a God or group of Gods, the non-existence of another life after death, and the creation of life by an "accident" of nature, so that it chooses to participate and become involved directly in community-based activities related to the major global problems described in this and earlier chapters, including overpopulation and family planning, vanishing food supplies, land-grabbing, destruction of ecological systems, insufficient water resources, vanishing seafood resources, and climate change. It is with pleasure that we acknowledge to have received inspiration from the column writings of *C.J. Werleman*[4] who also ardently feels that atheism must go beyond "glamorizing disbelief", and ought to support community causes and address inequality issues. Thus, the pillars and advocacies of Community-based Atheism (BCA) that we propose are as follows:

- A **God** or group of Gods do not exist, *never existed*.
- There is *no second life* after death.
- Life in this planet was created *by accident*, by a unique set of physical and chemical events, without the intervention of any supernatural entity.
- There is *no life in any other planet* or body in the Universe.
- There are two universes: (1) a first Universe of planets, stars, and galaxies out there, and (2) a second Universe of our species and its human emotions. The first Universe is *totally indifferent* to the second Universe.
- As a species we are *totally responsible* for our actions, without need to blame any other entity or species.

(PowerPoint Slide 20)

Furthermore, *we choose to have as our purpose of being* the following advocacies:

- To pay service to our communities through participation in activities which would *help prevent religion wars,* due to differences in religious beliefs, or the absence of these.

- To pay service to our communities through participation in activities which support *family planning,* contraception, and control of overpopulation.

- To pay service to our communities through participation in activities which would *prevent vanishing of land food resources* through protection of ecological sea systems.

- To pay service to our communities through participation in activities which would *prevent vanishing of seafood resources* through protection of ecological sea systems.

- To pay service to our communities through participation in activities which would prevent mass *land-grabbing* causing the displacement and un-empowerment of peoples in our planet.

- To pay service to our communities through participation in activities which would prevent the destruction of *ecological systems.*

- To pay service to our communities through participation in activities which support the availability and adequacy of *water resources* for peoples and land uses.

- To pay service to our communities through participation in activities which would promote the development of *alternatives sources of energy*, including *solar energy*.

- To pay service to our communities through participation in activities which would *prevent further climate change* through a number of programs, including control of CO_2 production and industrial pollution in general.

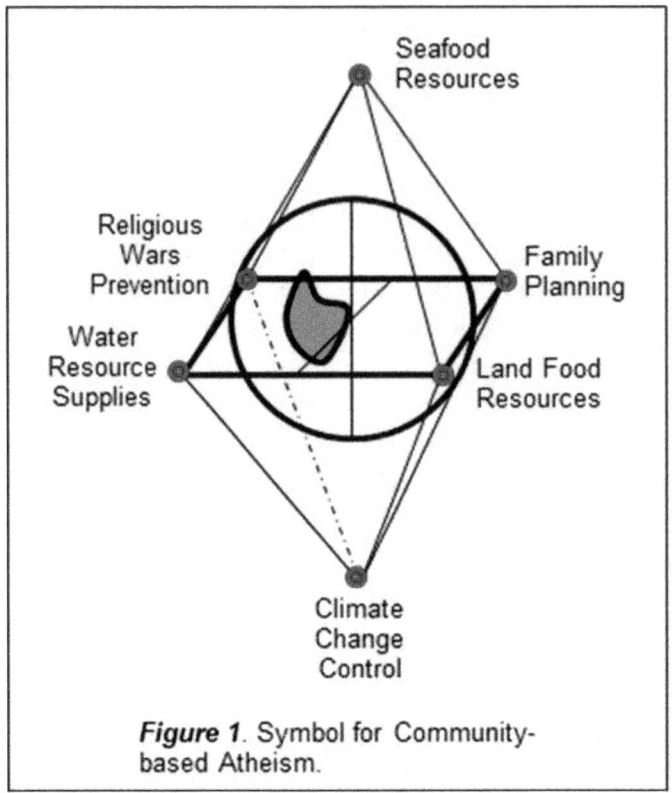

Figure 1. Symbol for Community-based Atheism.

It is proposed that the list of activities above in support of our communities include: (1) peaceful demonstrations, (2) involvement in the institutional system and legal system of countries involved, as well as in the corresponding systems of international organizations to either propose modification of existing laws or to propose new ones, (3) exhibits of textual and graphical material in private and public places in support of the above advocacies, (4) participation in community activities which would help communicate scientific evidence on the origin of our species, life in general, and the Universe, (5) respect for all religions, and (6) the promotion of values of respect, integrity, and service in our communities. Thank you.

The need for recognition in our social hierarchies

--I have another question, please. My question has to do with your statement about "social hierarchies", and how these hierarchies integrate into the contents of Scenario 8. Are you talking about "social models" in our communities?

--Good question. There is a belief among some of us in the IWP research team that basically says that human beings aspire to recognition in their own communities; a belief that says that even if we were to achieve economic and political balance in the planet, each one of us still wants some personal recognition and a hierarchical position in in our community, even if that position was to be merely temporal. Really? Each one of us aspires to personal recognition in his/her community? The hypothesis being here that if each one of us in the planet had a moment of personal recognition in his/her own community, there might be less of an independent personal effort to acquire such persona recognition by any other means. Fewer community conflicts, fewer nation rivalries, fewer wars? Possibly, I would like to postulate. To that effect let us consider *Maslow's Hierarchy of Needs.*

"What motivates people, what is their motive to do their work well and how can they be encouraged to perform even better?

To get a better understanding of this process, the psychologist *Abraham Maslow*[4] developed a H*ierarchy of Needs model in 1934*, in which he described five different levels of gratification of needs.

The hierarchy of needs is known as *Maslow Pyramid or theory of human behavior* and is still used in the corporate sector.

Levels of the Hierarchy of Needs

According to Abraham Maslow people are always motivated to satisfy their needs both at home and at work.

He does not make distinctions based on age. He categorized human needs into five hierarchical levels (Hierarchy of Needs).

He made the assumption that an advanced level can only be reached when the previous level of needs has been fulfilled.

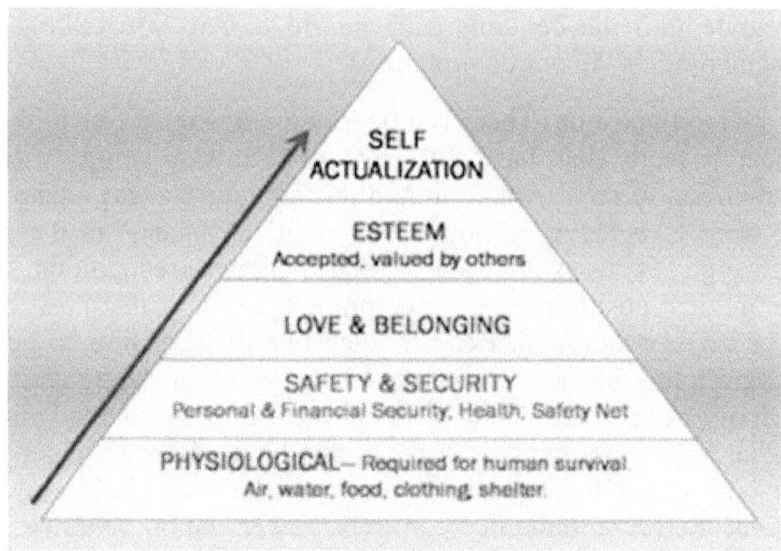

Figure 1. Pyramid of social needs in *Maslow's Hierarchy*. [1]

According to Abraham Maslow it is not possible to skip a level of the Hierarchy of Needs.

That is why it is important to fulfil the need that has been skipped or lost at a later date. The lowest level of Maslow's hierarchy of needs is the foundation of the pyramid. This is where the needs pattern begins.

These basic needs apply to everyone. The higher the level, the more difficult it becomes to satisfy the needs.

1. Physiological Needs

These include the most basic needs that are vital to survival, such as the need for food, water and sleep (**Primary Needs**).

Without fulfilment of these primary needs, people cannot function properly and they can fall ill.

Transition to work: A salary pays for most of these basic needs (food and drink).

2. Safety Needs

Every person wants security, safety and stability (*Secondary Needs*).

This can also be translated into peace, order and health.

This needs category also includes the security of a roof over one's head. If you make the transition to work: steady work, for instance a long-term contract, provides stability and security for the long term.

This ensures security with respect to housing and providing for the family.

3. Belonging

People are social beings and need social contacts. They wish to belong to a group. Friendship, acceptance, caring for other people and intimacy are important needs.

Transition to work: an employee will only invest time in social contacts on the work floor and be loyal to colleagues when they have been given the security of a long-term contract.

The employee now feels that they are more a part of the group.

4. Esteem

After investing in social contacts, people need esteem and recognition for what they do (*Recognition Needs*). Self-respect is crucial in this.

Only when these needs are met, they will need esteem, recognition and respect from other people.

Transition to the work place: The employer holds a motivator with this needs category.

Apart from the height of the salary, there are other factors that can motivate an employee.

As a result, compliments, trust and autonomy become important motivators for an employee.

5. Self-actualization

Because of a full development of certain qualities, this needs category will grow (**Development Needs**).

This can take place in different manners; from taking a course or night classes to taking on hobbies.

Transition to the workplace: Here too, motivators can be found. Some employees are extra stimulated if they are allowed to do certain courses or studies.

The incentive and appreciation for doing voluntary work, by offering a subsidy or leisure time, are part of this category.

Even when all the needs in the pyramid have been met, people will not be satisfied.

According to Abraham Maslow people will always have the urge to develop themselves and to chase after new needs, to be better at what they are good at.

A top sportsman wants to perform even better, an artist wants to pour more soul into his work and a manager wants to have an even bigger company.

This need is also called self-actualization.

Critical comments

(PowerPoint slide 31)

Today, Maslow's Hierarchy of Needs is heavily criticized by scientists.

There are situations in which it is not possible to substantiate the idea that these needs take place in a hierarchical order.

There are people who, despite very difficult circumstances, are perfectly capable of satisfying their social needs and who are capable of striving for recognition.

Furthermore, the various needs can merge with one another and they can vary from situation to situation.

In a workplace it is especially the younger employees who have a need for social contact, where as their older colleagues want to be recognized for their achievements.

As a third critical comment it could be said that Maslow's Hierarchy of Needs is rather static.

The needs of employees change and depend on time, the situation, experience and by comparing themselves with others.

Finally, *Clayton Alderfer* states in his ERG theory that people can regress to lower level needs despite the fact that these needs have already been fulfilled."

The same person raises his hand again. He appears skeptical.

--Yes, an interesting hierarchy of needs as proposed by Mr. Maslow, *but how does it relate to our concern for the negative impact of fratricide wars on our co-existence in our planet*?

--I was getting to that, so here we go. In fact, I will get right into an example to illustrate the point which our IWP research team would like to make. Consider the very recent terrorist attacks perpetrated initially in *France* by *ISIS* (Islamic State of Iran and Syria), the Jihadist militant group. Well, many of the perpetrators of such horrific and criminal attacks were born in France, in fact those men were first-generation and second-generation French citizens. So why did they commit such atrocities against their fellow human beings, why did they enter intro such horrible fratricide wars? Some members in our IWP team are of the opinion that those men had been ignored, did not count, that they found no place in their "social hierarchy." They parents had been immigrants, they were living in France, but they lacked "recognition" by the vast majority of the French population. They lived in their own neighborhoods with their traditional ways and culture, but without recognition by the other cultures in France. So when they were invited to join the ISIS organization they saw there an opportunity to participate in activities – terrorist and criminal activities – which they felt would place them on "first page" on radio, TV, and other mass media. They got one sort of recognition, for sure.

--OK, I get your point. But what do you propose we do in our communities relevant to this topic?

--There are many things which we could do in our communities, we believe. Currently, only a very few talented people occupy places of distinction in our communities, as is the case of teachers, lawyers, priests, rabbis, mullahs, medical doctors, City Hall personnel and related citizens. As such they obtain recognition in our "social hierarchies." But what happens to the other 98% of the population, working at home, rearing children, studying, and working for small companies? Where do they get "recognition" for their lives, their existence, their place in society?

In contrast, nowadays many communities in several countries give a money subsidy to emigrants in their communities, a "minimum wage" to unemployed citizens, etc. while remaining unable to participate in the life of their respective communities. Europe nowadays does just that: give money amounts to families of immigrants, and then just forget about them, their lives, their children, and their future in those communities.

We suggest that our "social models" be modified so as to allow activities and positions such as "City Mayor for one day", "local police officer for one week", "fireman for one week", "nurse in a residence for elderly people for one month", "member of the urban development committee for one month", and so forth. The idea being to give people in that 98% of the population the opportunity to participate in the social hierarchies of our communities. In time there would appear procedures, honors, prices, and money rewards to assign to individuals, and to recognize their presence, activity, and contribution to our societies. More recognition to citizens in their respective communities, less need for wars.

--So what is the *probability of this Scenario 8 happening*? – The question coming from still another person in the audience.

--Well, as we have done in earlier scenarios, we leave that question and response to our director in the IWP team. Dr. Finley, if you will, please?

--Thank you all, I'll try my best. This scenario would be highly beneficial to our human species on planet Earth, in my opinion.

However, I would only give it a 10%-15% probability of actually happening because it pretends to change the thinking of hundreds of Millions of people in our planet regarding that set of beliefs which we announced at the beginning of our presentation this afternoon. Many people go to wars and die thinking that an "afterlife" awaits them full of riches, sex, social recognition, and happiness in general. They go to war also after being convinced by others that "patriotism" is a highly justifiable reason for killing other human beings in "defense" of a nation, other human beings described as "different" when in reality they belong to our unique family of human beings originating in Africa, and leaving that continent on their way to the other continents 50,000 years ago. Human nature has no limits to what it will do to acquire property and power.

--Again, from our IWP research team, our thanks to everyone for being here today and, please, complete those one-page survey available on your tables.

Thank you all!

✳✳✳

NOTES

- **Bibliography**
- **Other books by this Author**
- **Biographical Note**

A note to the reader: Textual and graphical contents in this NOTES section are intended as an aid and service to the reader by expanding on topics and themes presented earlier in chapters throughout the book. Original sources, either books, articles, reports, or their *Uniform Resource Locator* (URLs; Web site address) in the Internet are cited. This author cannot vouch for the accuracy and/or validity of those materials in this "Internet age." It would not make sense either to ignore and not make use of the bountiful collections available on the Internet, gathered by thousands of individuals, from amateur writers, to published authors, and disciplined historians, just because these sources may not be "100% accurate, 100% of the time." Instead, it is proposed here, *it is the reader's responsibility* to determine if these materials are useful, accurate, valid, and sufficient for his/her personal use, and to decide if additional materials ought to be researched and gathered as well.

BILIOGRAPHY

Chapter 1: Introduction

None.

Chapter 2: The Global Community Today

[1]   *World   population   statistics,*   courtesy   of
https://en.wikipedia.org/wiki/World_population

[2] *Major Religions* by continent and country, courtesy of:
https://en.wikipedia.org/wiki/List_of_religious_populations

[3] *Origins of Homo Sapiens Sapiens,* courtesy of :
https://es.wikipedia.org/wiki/Homo_sapiens

[4] **Economic GDP** by country, courtesy of:
https://en.wikipedia.org/wiki/World_economy

[5] *Measures of Poverty*, courtesy of:
https://en.wikipedia.org/wiki/Measuring_poverty

Chapter 3: Formulation of Criteria
for the Eight Scenarios

[1] *Tree of life*, courtesy of
:http://en.wikipedia.org/wiki/Evolutionary_history_of_life

[2] *History of life on Earth*, diagram, courtesy of:
http://en.wikipedia.org/wiki/Evolutionary_history_of_life

[3] *"Pakistan, populations pressures"*, 1 April 2013, **World Population Awareness (WPA)**, in:
www.overpopulation.org/Asia

[4] *"Bangladesh: Land Scarcity and Rising Population"*,
10 March 2013, Financial Express Bangladesh, in:
www.overpopulation.org/Asia

[5] *"Family Planning Pilot Project in Philippines is a Success Story*
3 March 2013, by Bonnie Tillery, in:
www.overpopulation.org/Asia

[6] *"Fertility Rates Fall, but Global Population Explosion Goes on*
",
22 July 2012, Los Angeles Times, in:
www.overpopulation.org/Asia

[7] "Food Security Overview", 10 Oct 2014, World Bank Organization, in:
http://www.worldbank.org/en/topic/foodsecurity/overview#1

[8] *"Inside the Looming Food Crisis"*, extreme weather, booming populations make building the food supply a challenge, by Dennis Dimick, 22 May 2014, **National Geographic**, in:
http://news.nationalgeographic.com/news/2014/05/140522-food-crisis-vulnerable-weather-climate-future/

[9] *"Chinese firms and Gulf sheiks are snatching up farmland worldwide. Why?"*, by Brad Plumer, 26 January 2013, **National Geographic**, in: http://news.nationalgeographic.com/news/2014/05/140522-food-crisis-vulnerable-weather-climate-future/

[10] *"Seafood Crisis"*, **National Geographic**, October 2010, in http://ngm.nationalgeographic.com/2010/10/seafood-crisis/greenberg-text/1

Chapter 4: Weapon Technologies and Human Ailments

[1] Use of *chemical weapons* in history, courtesy of: en.wikipedia.org/wiki/chemical_warfare

[2] **List of States with Nuclear weapons**, courtesy of: https://en.wikipedia.org/wiki/List_of_states_with_nuclear_weapons

Chapter 5: Scenario 1: Overpopulation and Slow Death, Food Shortages, and Internal Conflict

None.

Chapter 6: Scenario 2: Guerrilla Warfare across the Planet

[3] **List of countries by Oil production**, courtesy of: https://en.wikipedia.org/wiki/List_of_countries_by_Oil_production

[4] **Oil production and use by Country**, courtesy of: https://www.bp.com/content/dam/bp/pdf/energy-economics/statistical-review-2016/bp-statistical-review-of-world-energy-2016-full-report.pdf

Chapter 7: Scenario 3: Climate Change and Environmental Holocaust

[1] *Environmental holocaust as a result of global warming,* courtesy of: https://feww.wordpress.com/tag/environmental-holocaust-2

[2] *"Estimating the global conservation status of more than 15,000 Amazonian tree species"*, courtesy of: http://advances.sciencemag.org/content/1/10/e1500936.full

[3] *"Trinity of Death: Contaminated Soil, Smog, and Polluted Water"*, courtesy of: https://feww.wordpress.com/2014/04/18/how-your-world-continued-shrinking/

[4] *"Apocaliptic Smog Covers 10% of China"*, courtesy of: https://feww.wordpress.com/2014/02/25/apocalyptic-smog-covers-10-percent-of-china/

[5] *"Massive leak occurs in a treatment system at Fukushima nuclear plant"*, courtesy of : https://feww.wordpress.com/2014/04/16/1-1-tons-of-radioactive-water-leaks-from-fukushima/

[6] *"Extremely high levels of air pollution are spreading across parts of England"*, courtesy of: https://feww.wordpress.com/2014/04/03/name-one-other-species/

[7] *Methane emissions 1,000 higher than EPA estimates,* courtesy of: https://feww.wordpress.com/2014/04/17/fracking-methane-emissions-hugely-underestimated-by-epa-study/

[13] *Climatic Change*, "Report: Climatic change Crisis Catastrophic", by Hilary Whiteman, London (CNN), in www.cnn.com/2009/world/europe/05/29/annan.climate.change.human

Chapter 8: Scenario 4: Nuclear Warfare and Death of the Human Genome

[1] *"Nuclear Holocaust"*, courtesy of: https://en.wikipedia.org/wiki/Nuclear_holocaust

[2] *"List of States with nuclear weapons"*, courtesy of: https://en.wikipedia.org/wiki/List_of_states_with_nuclear_weapons#Pakistan

[3] *Military exercises* to plan for a nuclear war, courtesy of: https://en.wikipedia.org/wiki/World_War_III

Chapter 9: Scenario 5: Diversity of Religions and Beliefs in the World

[2] Lapidus, Ira (2002). *A History of Islamic Societies* (2nd ed.). Cambridge University Press. ISBN 978-0-521-77933-3.

[3] Islam in China, in: http://www.bbc.co.uk/religion/religions/islam/history/china_1.shtml

[5] "The ancestors of today's primates were small and ate insects", **Figure 1**, Chapter 2, courtesy of **Boyd y Silk (2003)** and its publisher.

[6] The genetic variety of modern hominids, in Boyd and Silk (2003), pgs. 422-446.

[7] The human throat, Boyd and Silk (2003), pg. 450.

[8] Silvio's open space, Boyd and Silk (2003), pg. 461.

[10] *On Richard Leakey*, courtesy of: http://www.britannica.com/EBchecked/topic/333898/Richard-Leakey

[4] Sufi Order, in: Esposito, John (2004). *Islam: The Straight Path* (3rd Rev Upd ed.). Oxford University Press. ISBN 978-0-19-518266-8.

[12] Al-Khalili, Jim (2008-01-30). "It's time to herald the Arabic science that prefigure Darwin and Newton". *The Guardian*. London.

[6] Quran in English, courtesy of: http://qurango.com/english.html

[13] Islam expansion into Europe, in:
http://www.yale.edu/yup/pdf/cim6.pdf

[16] ***Women's rights within Islam***, courtesy of: Rippin, Andrew
(2001). *Muslims: Their Religious Beliefs and Practices* (2nd
ed.). Routledge. ISBN 978-0-415-21781-1, page 288.

[9] Expansion of Islam in modern times, courtesy of: Onishi,
Norimitsu (2001-11-01). "Rising Muslim power causes unrest
in Nigeria and elsewhere". *New York Times*.

[14] *"Muslims Say Their Faith Growing Fast in Africa"*
by Arthur Asiimwe and William Maclean (Reuters, November
14, 2004), in http://wwrn.org/articles/14286/?&place=eastern-
africa

[15] ***Shia, Shiites***, in:
http://www.britannica.com/EBchecked/topic/540503/Shiite

Chapter 10: Scenario 6: Promote Economic
and Political Balance across the Global Community

[1] ***Economic Goals*** of the United Nations (UN), courtesy of:
www.un.org/sustainable development/

[2] ***"Globalization: A Brief Overview"***, by International
monetary Fund, May 2008, courtesy of:
https://www.imf.org/external/np/exr/ib/2008/053008.htm

[3] *"A Multi-Polar World"*, by Scott Webster, 24 October 2013,
courtesy of: http://scottwebsterministries.org/article/a-multi-
polar-world

Chapter 11: Search, Find,
and Populate other Planets

[1] *"Space colonization"*, January 2017, courtesy of:
https://en.wikipedia.org/wiki/Space_colonization

[2] ***"Seven Earth-like planets discovered orbiting a star"***,
courtesy of: http://edition.cnn.com/2017/02/22/world/new-
exoplanets-discovery-nasa/

[3]   *"What It Will Take to Become an Interstellar Civilization"*, by Sarah Scoles, October 28, 2014, courtesy of: www.blogs.discovermagazine/crux/2014/10/28/takes-interstellar-civilization

[4]   *Space colonization*, courtesy of: www.wikipedia/wiki/space_colonization_objections

[5]   *"Coming Age of Space Colonization"*, 20 March 2013, courtesy of: https://www.theatlantic.com/technology/archive/2013/03/the-coming-age-of-space-colonization/273818/

[6]   *"Biosphere 2"*, courtesy of: https://en.wikipedia.org/wiki/Biosphere_2

Chapter 12: <u>Scenario 8:</u> Coming to terms with our Human Nature and Place in the Universe

[1]   *"It is already too late to save Humanity"*, by Mike Adams, 19 June 2015, courtesy of: http://www.naturalnews.com/050127_human_extinction_Professor_Frank_Fenner_population_control.html

[2]   *The Gaia hypothesis*, by James Loveluck, May 2017, courtesy of: http://europe.newsweek.com/james-lovelock-saving-planet-foolish-romantic-extravagance-327941

[3]   *"To deceive is an advantage in the evolution process"*, by American scientist **Robert Trivers**, in his book The Folly of Fools, 11 May 2012, in www.muyinterestante.es

[4]   *"Maslow's theory on social hierarchy"*, courtesy of: https://simplypsychology.org/maslow.html

Other Books by this Author [Return]

A short list:

(1) *Planet Earth at its Limits: Overpopulation, Limited Resources, Religious Wars, and Climatic Change...* Euskal Herria 21st Century publisher, Arrasate, Basque Country, February 2013. Available in www.amazon.com

(2) *Frontiers in DNA Research: Questions and Answers about your own DNA*, 300 pages...ISBN-13 978-1542766029... Euskal Herria 21st Century publisher, Arrasate, Basque Country, February 2013. Available in www.amazon.com

(3)

(4) *Book: Speak Up! A Practical Guide to Modern English* (20 Secrets to successful pronunciation and communication in English), 239 pages, Euskal Herria 21st Century publisher, Arrasate, Basque Country, February 2013. Available in www.amazon.com

(5) *Book: Enterprise Architectures and Digital Administration: Planning, Design, and Assessment (Arquitecturas Empresariales y Administracion Digital: Planificacion, Diseño, y Asesoría)*, World Scientific Press, New York, 565 pages, Abril 2007 (See book promotion: http://www.worldscibooks.com/business/6239.html).

(6) *Book: Euskal Herria Estado-Nacion en el Siglo 21: Una Nueva Arquitectura Socio-Politica*, <u>Editorial Euskal Herria 21st Century</u>, 487 pages, February 2008 (See book promotion by distributor <u>www.elkar.es</u>)

(7) *Book: "When the Parallel Worlds Co-Existed"*, Book 1 of the Novel-Trilogy "Women, the New Architects of Society", publisher: <u>Euskal Herria 21st Century</u>, ISBN 978 846 147 5544, February 2011.

(8) *Book: "The Pope's Red Shoes" (Los Zapatos Rojos del Papa)*, libro 2 of the Novel-Trilogy "Women, the New Architects of Society", publisher: <u>Euskal Herria 21st Century</u>, ISBN 978-84-614-7551-3, February 2012.

(9) *"La Brujas de Zugarramurdi: El Musical"* (in Ingles, Euskera, and Castellano), 4 acts, 20 scenes, 45 songs, 55 pages, Editorial Euskal Herria 21st Century, Arrasate, Gipuzkoa, Basque Country, 2012.

(10) *Article: "Findings of a Basque-American in Euskal Herria Today: Betrayal, Reality, and the Winds of Change"*, published in the *Journal of the Society of Basque Studies in America*, Vol. XXVIII, pages 42-64, 2008.

(11) *"Witches and Wizards who never existed" (Brujas y Brujos que no existieron: Auto de Fe de 1610 en Logroño)* (wit Aloña Altuna), 48 pages, in **AUNIA**, No. 32, 2011.

All these books and articles are available in the INTERNET, in <u>www.amazon.com</u>. In that web site simply type in this author's name and the list of books above appear. The reader can read several chapters free of charge. Book prices are very reasonable.

NOTES

For more information, please contact this author at:
agoikoetxea1@telefonica.net

Mila esker!
Gracias!
Thank you!

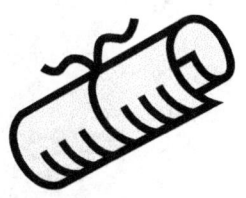

Biographical Note about this Author [Return]

Author

Ambrose Goikoetxea Martínez, Ph.D. (Biasteri-Laguardia, Alava, Basque Country, 1952-)

This Basque-American author was born in the town of ***Laguardia-Biasteri*** ("La Guardia of Nafarroa") province of Alaba, Euskal Herria (Basque Country) in a family of carpenters on his father's side (the *Goikoetxeas*), farmers, monks and nuns on his mother's side (the *Martinez*). At the age of 5, and right after the Spanish Civil War, the family migrated to the United States in search of a better life. B.S. (Bachelor of Science) in Aeronautical Engineering, California Polytechnic Institute (Cal-Poly), Pomona, California, 1969; M.S. (Master of Sciences), Mechanical Engineering,  University of California at Los Angeles, 1970; ***Doctor of Philosophy (Ph.D.) in Systems Engineering and Economics, University of Arizona, Tucson, 1977, USA***. He went on to work as a professor at Oklahoma State University (OSU), and George Washington University (GWU), and as an engineer at several corporations in the USA, for the next 40 years until 2004 when he returned from his experience in the "Basque Diaspora" to Euskal Herria to continue his work, this time applying his skills and experience to social needs in Basque society. A brother of his is today a medical doctor in San Diego, California, a sister is an English teacher and social worker at Simi Valley, California, a son is a medical doctor, an AIDS researcher at the University of San Diego, California, and a second son is a software engineer and computer systems architect in Boston, Massachusetts, USA.

Currently a "half-time" retired person, he is a part-time English teacher at several local academies, including *Mondragon Lingua* (since 2007) with base in Arrasate-Mondragon, *Espolon Academy* with base in Bergara, and *Elduaien Academy* with base in Tolosa. He is of the opinion that both mind and body must be exercised in order to stay healthy. Towards that end, teaching English gives him the opportunity to share his diverse experience in the Anglo-Saxon world with hard-working and dedicated people in the Basque Country and Spain, to learn of new initiatives and ventures in the business world, and to work with students, other teachers, and business managers. Additionally, he feels these are very exciting times as the new information technologies in the Internet and specialized software tools offer great capabilities that can be used to improve pronunciation and communication skills in English as a second language. It's a whole new world, as he says; the opportunities in English teaching and learning are boundless.

Prior to teaching English, his teaching and research activities have been many and varied in the USA, Latin America, Basque Country, and Spain. Over 20 years of experience in the design and application of systems engineering principles and methodological frameworks to problems in computer architecture design, decision support systems, database design, transportation systems, data fusion for military object classification, design of multiple criteria decision support systems (MCDSS), and programmatic efficiency in city hall, public administrations, local and regional governments, and military services (US Air Force, Army, Navy, and Coast Guard).

Founder and Director (part-time) of the *Euskal Herria 21st Century Foundation*, based in Arrasate-Mondragon since February

2007, an organization that works with citizens to re-build the social and political fiber in Euskal Herria (Basque Country); bilateral projects with Eusko Etxeak and other Basque centers in the USA and Latin America (web site: www.euskalherriasiglo21.org). From 2004 to 2007 he worked in the Department of Information Sciences at the *University of Mondragon (MU)* where he taught graduate and undergraduate courses in software engineering, design of software processes with UML (Unified Modelling Language) tools, enterprise computer architectures, and digital government administration. There, as Director of the *e-Democracy Project*, his team lent technical support to members of the Basque Parliament in an assessment of the use of the new information technologies (e.g., e-mail, digital TV, electronic signature, etc.) in 74 Parliaments and regions with legislative capacity in the European Union (EU).

Earlier, 1999-2004, he was a Sr. Information Systems Engineer at the Center for Excellence in Software Engineering (W908) of the *MITRE Corporation*, Virginia, EE.UU. In that capacity he supported business systems architecture development in the IRS Modernization Enterprise Architecture, applications test planning and business rules integration in the Customer Account Data Engine (CADE), and performance assessment of messaging middleware (MQSeries). At MITRE he also completed an architecture design to support intra-site and inter-site data backup, failover, and recovery for the Army's Defense Message System (DMS). Prior to joining MITRE he was the Performance and Capacity lead engineer for the Global Transportation Network (GTN) at Lockheed Martin Corporation, in Manassas, Virginia (1997-1999) where he created the system performance and capacity planning (PCP) group, designed and instituted PCP processes, responsible for setting up a suite of modelling and measurement tools.

An Associate Professor in the Systems Engineering Department, George Mason University, 1985-1999. President and Technical Director of Integrated Technologies and Research, Inc., from 1995 to 1997, where he designed and developed decision support systems for the U.S. Army Corps of Engineers. In 1985 and 1986 he was NASA-ASEE Research Fellow at the Goddard Space

Flight Center, in Greenbelt, Maryland, designing decision support systems for NASA managers to assist with systems engineering and configuration functions of space projects; also, a member of the Man-Machine Interface Design Group. 1979, NASA-ASEE Research Fellow, Jet Propulsion Laboratory of the California Institute of Technology (Cal-Tech); evaluation and selection of projects in the areas of solar-thermo power plants, underground nuclear plant location analysis, and urban public transportation systems.

Organizer and **General Chair** of the *IX-th International Conference on Multiple Criteria Decision Making (MCDM), Fairfax, Virginia,* August 5-8, 1990. Dr. Goikoetxea is a recognized speaker at international conferences on system performance modelling, decision analysis, distributed database design, and risk analysis. A member of several engineering professional organizations, and a lecturer in the Department of Engineering Management, Department of Operations Research, and the Department of Management Science of George Washington University, 1990-2008.

As elected **Program Chair** of the *Association for Development of the Information Society* (IADIS), Dr. Goikoetxea brought together Mondragon University and Universidade Alberta of Portugal to celebrate an international conference on Applied Computing and Web-Based Communities (see www.iadis.org/ac2006, www.aidis.org/wbc2006) in San Sebastian, Basque Country, 25-28 February 2006 with the participation of 250-275 persons from 25 countries.

A member of ***Iniciativa Atea*** *(Atheist Initiative)*, a non-profit international organization with base in Iruña-Pamplona, Basque Country, which contributes to free-thinking and cultural initiatives for the benefit of citizens in general. A Voluntary in the **Green Peace** organization in support of pro-environment initiatives, a member of various organizations that support and actively promote **women's human rights**. A avid bike rider, enthusiastic cook, and weight lifter (3 times a week at a local gym), he now lives with his wife Aloña in Arrasate, Gipuzkoa, Basque Country, and often travels to the town of Laguardia-Biasteri, Alava, Boston, Massachusetts, San Diego, California, and Boise, Idaho, USA, to visit relatives and friends.

***